Marxist Introductions

General Editor
Steven Lukes

D0144080

Marxism and Morality

STEVEN LUKES

Oxford New York
OXFORD UNIVERSITY PRESS

Oxford University Press, Walton Street, Oxford OX2 6DP

Oxford New York Toronto
Delhi Bombay Calcutta Madras Karachi
Petaling Jaya Singapore Hong Kong Tokyo
Nairobi Dar es Salaam Cape Town
Melbourne Auckland

and associated companies in
Berlin Ibadan

Oxford is a trade mark of Oxford University Press

First published 1985
First issued as an Oxford University Press paperback 1987
Paperback reprinted 1988

British Library Cataloguing in Publication Data
Lukes, Steven
Marxism and morality.—(Marxist introductions).
1. Communism 2. Socialism
I. Title II. Series
335.4 HX73
ISBN 0–19–282074–5

Library of Congress Cataloging in Publication Data
Lukes, Steven.
Marxism and morality.
(Marxist introductions)
Bibliography: p. Includes index.
1. Communist ethics—History. 2. Marx, Karl,
1818–1883—Ethics. 3. Engels, Friedrich, 1820–1895—
Ethics. I. Title. II. Series.
[BJ1390.L79 1987] 171'.7 87–15292
ISBN 0–19–282074–5 (pbk.)

Printed in Great Britain by
The Guernsey Press Co. Ltd.
Guernsey, Channel Islands

To the memory of Peter Sedgwick

With whom would the just man not sit
To help justice?
What medicine is too bitter
For the man who's dying?
What vileness should you not suffer to
Annihilate vileness?
If at last you could change the world, what
Could make you too good to do so?
Who are you?
Sink in filth
Embrace the butcher but
Change the world: it needs it!

(Brecht 1929–30: 25)

You who will emerge from the flood
In which we have gone under
Remember
When you speak of our failings
The dark time too
Which you have escaped.

For we went, changing countries oftener than our shoes
Through the wars of the classes, despairing
When there was injustice only, and no rebellion.

And yet we know:
Hatred, even of meanness
Contorts the features.
Anger, even against injustice
Makes the voice hoarse. Oh, we
Who wanted to prepare the ground for friendliness
Could not ourselves be friendly.

But you, when the time comes at last
And man is a helper to man
Think of us
With forbearance.

(Brecht 1938: 319–20)

With the rest of my generation I firmly believed that the ends justified the means. Our great goal was the universal triumph of Communism, and for the sake of that goal everything was permissible—to lie, to steal, to destroy hundreds of thousands and even millions of people, all those who were hindering our work or could hinder it, everyone who stood in the way. And to hesitate or doubt about all this was to give in to 'intellectual squeamishness' and 'stupid liberalism', the attributes of people who 'could not see the forest for the trees.'

That was how I reasoned, and everyone like me, even when I did have my doubts, when I believed what Trotsky and Bukharin were saying, when I saw what 'total collectivization' meant—how they 'kulakized' and 'dekulakized', how mercilessly they stripped the peasants in the winter of 1932–3. I took part in this myself, scouring the countryside, searching for hidden grain, testing the earth with an iron rod for loose spots that might lead to buried grain. With the others, I emptied out the old folks' storage chests, stopping my ears to the children's crying and the women's wails. For I was convinced that I was accomplishing the great and necessary transformation of the countryside; that in the days to come the people who lived there would be better off for it; that their distress and suffering were a result of their own ignorance or the machinations of the class enemy; that those who sent me—and I myself—knew better than the peasants how they should live, what they should sow and when they should plough.

In the terrible spring of 1933 I saw people dying from hunger. I saw women and children with distended bellies, turning blue, still breathing but with vacant, lifeless eyes. And corpses—corpses in ragged sheepskin coats and cheap felt boots; corpses in peasant huts, in the melting snow of old Vologda, under the bridges of Kharkov . . . I saw all this and did not go out of my mind or commit suicide. Nor did I curse those who had sent me to take away the peasants' grain in the winter, and in the spring to persuade the barely walking, skeleton-thin or sickly-swollen people to go into the fields in order to 'fulfil the Bolshevik sowing plan in shock-worker style'.

Nor did I lose my faith. As before, I believed because I wanted to believe. Thus from time immemorial men have believed when possessed by a desire to serve powers and values above and beyond humanity: gods, emperors, states; ideals of virtue, freedom, nation, race, class, party. . . .

Any single-minded attempt to realize these ideals exacts its toll of human sacrifice. In the name of the noblest visions promising eternal happiness to their descendants, such men bring merciless ruin on their contemporaries. Bestowing paradise on the dead, they maim and destroy the living. They become unprincipled liars and unrelenting executioners, all the while seeing themselves as virtuous and honourable militants—convinced that if they are forced into villainy, it is for the sake of future good, and that if they have to lie, it is in the name of eternal truths.

> *Und willst du nicht mein Bruder sein*
> *So schlag ich dir dein Schädel ein.*
> [And if you won't be my brother
> I'll crack your skull open.].

they sing in a Landsknecht song.

That was how we thought and acted—we, the fanatical disciples of the all-saving ideals of Communism. When we saw the base and cruel acts that were committed in the name of our exalted notions of good, and when we ourselves took part in those actions, what we feared most was to lose our heads, fall into doubt or heresy and forfeit our unbounded faith.

I was appalled by what I saw in the 1930s and was overcome by depression. But I would still my doubts the way I had learned to: 'we made a mistake', 'we went too far', 'we didn't take into consideration', 'the logic of the class struggle', 'objective historical need', 'using barbaric means to combat barbarism'. . . .

Good and evil, humanity and inhumanity—these seemed empty abstractions. I did not trouble myself with why 'humanity' should be abstract but 'historical necessity' and 'class consciousness' should be concrete. The concepts of conscience, honour, humaneness we dismissed as idealistic prejudices, 'intellectual' or 'bourgeois' and, hence, perverse.

(Kopelev 1975: 32–4)

Preface

In this book I am concerned with three questions. The first concerns marxism as a theory: what does it have to say about morality, and what answers does it offer to such central moral questions as 'What is justice?', 'Do people have rights and, if so, what rights do they have?', 'In what does the human good consist?', 'What means may we employ in the pursuit of our ends?'? The second concerns marxism in practice: what can now be said about the moral record of marxism as a social movement and as a system of rule, whether measured against its own values and standards or against others that may be thought appropriate? The third question concerns the relation between the first two, between marxism in theory and marxism in practice: what bearing has marxism's approach to morality and moral questions had upon its moral record in the struggle for and exercise of power? And conversely, what lessons may be drawn from an examination of that record with respect to marxism as a system of belief? Within the span of this book, I can only examine the first of these questions in any detail, but in doing so, I shall suggest my answers to the second and third. At the least, I hope that this book will both encourage and help others to face them, for they should be faced rather than either avoided or assumed to have self-evident answers.

To speak somewhat more bluntly, my purpose is to raise (I obviously cannot answer) the question whether the theory constructed by Marx and Engels, and developed by their successors, can in any respect and to any degree account for the moral disasters of marxism in practice. The record has, of course, been a mixed one. Marxists have an honourable place in the annals of resistance to oppression, above all to fascism and nazism; and in simple utilitarian terms, in developing societies, the achievements of marxists in power must be set against the ravages of death, hunger, unemployment, poverty, and disease often permitted and the tyranny and repression often practised by the historically available alternatives. The record must also be set against the human costs of capitalism, both within its heartlands and throughout its dependent periphery, and against the entire history of capitalist imperialism and neo-colonialism, whose massive endorsement of local brutalities, and suppression of individual and collective liberties are presently visible behind the moralistic façade of United States foreign policy, especially in South-East Asia and now in Central America. It should also be recalled that, as Trotsky remarked and Barrington Moore has shown, bourgeois or liberal democracy 'came into the world not at all through the democratic road' (Moore 1967, Trotsky 1938: 24). None the less, disasters there have been—above all, when they are measured, as they demand to be, against marxism's moral promise. Of course, many plausible explanations have been and can be offered (including the contributions of anti-socialist forces)—for the Bolshevik terror during and after the Civil War, for Stalin's terror, the purges and trials, the mass deportations and the vast network of labour camps, for the social catastrophe of Mao's Cultural Revolution, for the 'murderous utopia' of Pol Pot's Cambodia, and for the grim, surveillance-minded, demoralized world of contemporary 'actually existing socialism', above all in the USSR and Eastern Europe, where civil society and public life have been destroyed, and both marxist and moral vocabulary have become wholly devalued, the worthless currency of an empty rhetoric. The ironic culmination of these developments has been the general annihilation in such societies and beyond of the belief that the socialist project is worthy of allegiance, or even serious attention.

Those who derive satisfaction from this outcome may welcome this book as grist to their mill; others, for that very reason, may condemn it, or its publication, as unhelpful at a time when socialism needs all the friends it can muster (I write in the second year of Mrs Thatcher's second term). But *'pas d'ennemis à gauche'* has always been a dubious principle, stifling critical discussion; and it is no service to the cause of socialism to avoid meeting *its* enemies' strongest challenge. This book is, emphatically, not just another anti-marxist tract. Rather, it advances a hypothesis, that is both conceptual and historical, about the links between the marxist ethic and the spirit of socialism, in the belief that the latter can only flourish when some of these links have been severed. It is, in short, an attempt to come to grips with what is wrong with marxism's approach to the central questions of how we should and might live, in the conviction that this has borne some relation to what has gone wrong in practice.

That relation is not, however, to be captured by the metaphor of the germinating seed: this falsely and naïvely suggests that the historical developments were inherent in the theory, which could only come to fruition in these disastrous ways. Theories are plundered and their ideas selected and interpreted by historical actors pursuing their interests within objective conditions and under pressure of historical contingencies: it is all of these, in combination, not simply the logic of the theories, which explain historical outcomes. That much, at the very least, marxism has taught us. A better image is that of disablement: despite its many strengths, the theory of the founders was blind and deaf to, and silent about, certain ranges of moral questions— roughly, those concerning justice and rights, which set constraints on how people are to be treated in the here and now, and in the immediate future. This disability has been transmitted from the original theory to its main descendants, as Chapters Two and Six seek to show. It has also, I believe, characterized marxist ideology far and wide, though I do not show that here, or chart the influence it has had on the attitudes and conduct of marxists, prominent and obscure. But only the most vulgar of deterministic marxists could suppose such influence to be negligible.

Can these congenital defects of marxist theory be cured, and, if so, can this be done from within marxism, perhaps by developing lines of thought suggested but undeveloped by Marx himself? Possibly. My concerns in this book, however, are diagnostic only: with the structure of what has been marxism's distinctive approach to morality and moral questions, with its underlying rationale, and its possible consequences.

Acknowledgements

The argument of this book has grown slowly into its present shape. Essential to that process was the reaction of a group of friends in Prague, deeply and immovably sceptical of my very title, and a comment of Jerry Cohen's that it sounded like a short book. (It is a short book). No less essential were lively discussions with Denis Wrong and Tracy Strong, with each of whom I taught a course on this theme, and the invaluable reactions and suggestions of Anthony Arblaster, Tom Bottomore, David Caute, G. A. Cohen, Bill Connolly, Jon Elster, Tim Garton Ash, Norman Geras, Tony Giddens, David Held, Irving Howe, Wayne Hudson, Brian Knei-Paz, Yaakov Malkin, Alan Montefiore, Gary Ostrolenk, Frank Parkin, Jo Raz, Raphael Samuel, the late Peter Sedgwick, Nina Stanger, Christine Sypnowich, and Erik Olin Wright.

Contents

1 The Paradox Stated

Marxism's attitude to morality is paradoxical. A paradox, according to the *Oxford English Dictionary*, is a 'statement seemingly self-contradictory and absurd, though possibly well-founded and essentially true'. The aim of this chapter is to suggest that the traditional marxist view of morality is, indeed, paradoxical: a mixture of positions in apparent contradiction or at least tension with one another. The following chapter seeks to substantiate this suggestion by surveying what the marxist tradition has had to say on the topic. In Chapter 3, I shall try to resolve the paradox by showing that the view in question is only *seemingly* self-contradictory: that marxism offers a consistent and distinctive approach to morality and moral questions. I shall not, however, suggest that this approach is, as a whole, either 'well-founded' or 'essentially true'. The following two chapters illustrate and explore that approach. Focusing more directly on Marx's own thought, they consider in turn his contrasting treatments of two different domains of morality and ranges of moral questions. The final chapter asks whether, if this account is correct, marxism has anything distinctively moral to say in answer to Lenin's question: 'What is to be done?' and to its no less important corollary: 'What is not to be done?'

One preliminary definitional question: what do I mean by 'marxism' and by 'morality'? As to the first, we must, of course, bear in mind Marx's own dismissal of the label 'marxist'—'All I know', he used to say, 'is that I am not a marxist' (Engels 1890: 496)—and the great diversity of subsequent marxist sub-traditions. Nevertheless, my argument will be that, whatever other issues may have divided them, there has always been a certain coherent view that united very many, though not all, self-proclaimed marxists (and certainly all those who have been influential in practice) with Marx himself and his close collaborator Engels. It is that view which this book aims to articulate and analyze. I am, of course, aware that Marx's thought is remarkably rich, complex, and many-sided, expressed at different levels, in different contexts (sometimes polemical, sometimes journalistic, sometimes scientific), and in different tones of voice (sometimes ironic, sometimes demagogic, sometimes analytic, sometimes prophetic), and in many respects undeveloped and open-ended. I am also aware that the marxist tradition is no monolithic unity, but a contested terrain in which the solemn orthodoxies of the Second and Third Internationals have faced many and various forms of heterodoxy and revisionism, from Bernstein and the Austro-marxists to the Frankfurt and Budapest Schools. In speaking of 'marxism' in theory and practice, I do not mean to endorse any particular line of descent from Marx as legitimate, least of all that which runs via Engels, Plekhanov, and Lenin through dialectical materialism to Soviet-style communism. Orthodox 'scientific' and Russified marxism is only one line of (arrested) development within the tradition, which happens to have been the one that has had momentous world-historical effects in practice; and other lines have undoubtedly been far truer to the letter and spirit of Marx's thought. Rather, my claim is that there is a central structure of thinking, developed by Marx and Engels, which I seek here to exhibit, that has been partly constitutive of that tradition, and by which even the heterodox have been deeply imprinted. By that claim my entire argument stands or falls. I shall, when describing Marx's own views and writings, use the adjective 'marxian', and when describing those of his followers the adjective 'marxist'. I do not, however, intend these terms of art to imply, in general, that the marxist contradicts or even diverges from the marxian: where it does, I shall say so.

As for what 'morality' means, that will, I hope, become clear in the course of the analysis; let it suffice here to say that it concerns at least the domain of the right and the good, and questions of obligation, duty, fairness, virtue, character, the nature of the good life and the good society, and, behind these, assumptions about the nature of man, the preconditions for social life, the limits of its possible transformation, and the grounds of practical judgement.

The paradox in marxism's view of morality lies in the fact

that one set of positions central to marxism throughout its history when set beside another set of positions no less central appears to generate a striking contradiction.

On the one hand it is claimed that morality is a form of ideology, and thus social in origin, illusory in content, and serving class interests; that any given morality arises out of a particular stage in the development of the productive forces and relations and is relative to a particular mode of production and particular class interests; that there are no objective truths or eternal principles of morality; that the very form of morality, and general ideas such as freedom and justice that are 'common to all states of society', cannot 'completely vanish except with the total disappearance of class antagonisms' (Marx and Engels 1848: 504); that the proletarian sees morality, along with law and religion, as 'so many bourgeois prejudices, behind which lurk in ambush just as many bourgeois interests' (ibid.: 494–5); that marxism is opposed to all moralizing and rejects as out of date all moral vocabulary, and that the marxist critique of both capitalism and political economy is not moral but scientific.

On the other hand, no one can fail to notice that Marx's and marxist writings abound in moral judgements, implicit and explicit. From his earliest writings, where Marx expresses his hatred of servility, through the critique of alienation and the fragmentary visions of communism in the Paris Manuscripts and *The German Ideology*, to the excoriating attacks on factory conditions and the effects of exploitation in *Capital*, it is plain that Marx was fired by outrage and indignation and the burning desire for a better world that it is hard not to see as moral. The same applies to Engels, author of *The Condition of the Working Class in England*, a work full of moral criticism of the social conditions created by advancing industrial capitalism, which remained basic to his thought, and which Marx explicitly endorsed for its depiction of the 'moral degradation caused by the capitalistic exploitation of women and children' (Marx 1867: 399–400). The same applies to their followers down to the present day. Open practically any marxist text, however aseptically scientific or academic, and you will find condemnation, exhortation, and the vision of a better world. As for the socialist leaders, as Irving Howe has well said, few 'were of proletarian origin, few acted out of direct class needs, and most were inspired by moral visions their ideology somehow inhibited them from expressing' (Howe 1981: 492).

Notice that the paradox, the seeming contradiction, lies at the level of general belief. On the one hand, morality, as such, is explained, unmasked, and condemned as an anachronism; on the other, it is believed in and appealed to, and indeed urged upon others as relevant to political campaigns and struggles. I am not referring to a contrast between, say, bourgeois morality on the one hand, and authentic proletarian morality on the other. As we shall see, marxists have sometimes drawn this distinction, but it is no contradiction. I am concerned rather with marxist beliefs about morality and moral judgement *per se* and in general: in the absence of further explanation, these certainly do look contradictory. Nor am I referring to the contrast between theory and practice, between what marxists say and what they do. Often, where this contrast exists, it does constitute a contradiction, but it is one common to all political ideologies and creeds, and it is no paradox. In short, what is striking about marxism is its apparent commitment to both the rejection and the adoption of moral criticism and exhortation.

2   The Paradox Illustrated

In *The German Ideology*, Marx and Engels wrote of 'Morality, religion, metaphysics, and all the rest of ideology as well as the forms of consciousness corresponding to these', that they 'no longer retain the semblance of independence. They have no history, no development; but men, developing their material production and their material intercourse, alter, along with this their actual world, also their thinking and the products of their thinking' (Marx and Engels 1845–6: 36–7). The communists, they wrote,

do not preach *morality* at all, as Stirner does so extensively. They do not put to people the moral demand: love one another, do not be egoists, etc.; on the contrary, they are very well aware that egoism, just as much as selflessness, is in definite circumstances a necessary form of the self-assertion of individuals. (Ibid.: 247.)

Not only is morality dependent on and relative to changing material circumstances; it is also an illusion to be unmasked and a set of prejudices behind which class interests 'lurk in ambush' (Marx and Engels 1848: 495). The nature of the danger is clear from Marx's and Engels's attacks on the moral vocabulary of the 'True Socialists', foreshadowing later attacks by Kautsky, Luxemburg, and Lenin on 'Ethical Socialists':

In reality, the actual property owners stand on one side and the propertyless communist proletarians on the other. This opposition becomes keener day by day and is rapidly driving to a crisis. If, then, the theoretical representatives of the proletariat wish their literary activity to have any practical effect, they must first and foremost insist that all phrases are dropped which tend to dim the realisation of the sharpness of this opposition, all phrases which tend to conceal this opposition and may even give the bourgeoisie a chance to approach the communists for safety's sake on the strength of their philanthropic enthusiasms . . . it is . . . necessary to resist all phrases which obscure and dilute still further the realisation that communism is totally opposed to the existing world order. (Marx and Engels 1845–6: 469.)

As for communism, it was 'not for us a *state of affairs* which is to be established, an *ideal* to which reality [will] have to adjust itself. We call communism the *real* movement which abolishes the present state of things, the conditions of this movement result from the now existing premises' (ibid.: 49). Moral thinking stemmed, in effect, from a cognitive inadequacy that was itself historically determined and only surmountable at a certain stage of historical development:

It was only possible to discover the connection between the kinds of enjoyment open to individuals at any particular time and the class relations in which they live, and the conditions of production and intercourse which give rise to these relations, the narrowness of the hitherto existing forms of enjoyment, which were outside the actual content of the life of people and in contradiction to it, the connection between every philosophy of enjoyment and the enjoyment actually present and the hypocrisy of such a philosophy which treated all individuals without distinction—it was, of course, only possible to discover all this when it became possible to criticise the conditions of production and intercourse in the hitherto-existing world, that is when the contradiction between the bourgeoisie and proletariat had given rise to communist and socialist views. That shattered the basis of all morality, whether the morality of asceticism or of enjoyment. (Ibid.: 418–19.)

Marx maintained these positions throughout his life. It is true that in 1864 he helped draft the General Rules of the International Working Men's Association, whose members were enjoined to acknowledge 'truth, justice, and morality, as the basis of their conduct towards each other and towards all men, without regard to colour, creed or nationality', and the principle of 'no rights without duties, no duties without rights'; while the struggle for the emancipation of the working classes is described as a struggle for 'equal rights and duties, and the abolition of all class rule' (Marx 1864a: 386–9). Moreover, in his Inaugural Address, Marx urged workers to 'vindicate the simple laws of morals and justice, which ought to govern the relations of private individuals, as the rules paramount of the intercourse of nations' (Marx 1864b: 385). On the other hand, he explained these unfortunate phrases in a letter to Engels of 4 November 1864: 'I was obliged', he wrote, 'to insert two phrases about "duty" and "right" into the Preamble to the

Rules, ditto "truth, morality and justice" but these are placed in such a way that they can do no harm' (Marx 1864c: 182). As he had written two decades earlier, the appeal to the rights of the workers was 'only a means of making them take shape as "they", as a revolutionary united mass' (Marx and Engels 1845–6: 323).

In *Capital*, Marx scorns Proudhon's appeal to an ideal of justice. What opinion, he asks,

should we have of a chemist, who, instead of studying the actual laws of the molecular changes in the composition and decomposition of matter, and on that foundation solving definite problems, claimed to regulate the composition and decomposition of matter by means of 'eternal ideas', of *'naturalité'* and *'affinité'*? Do we really know any more about 'usury' when we say it contradicts *'justice éternelle'*, *'équité éternelle'*, *'mutualité éternelle'*, and other *'vérités éternelles'* than the fathers of the church did when they said it was incompatible with *'grace éternelle'*, *'foi éternelle'* and *'la volonté éternelle de Dieu'*? (Marx 1867: 84–5.)

In the *Critique of the Gotha Programme*, he once more makes clear his rejection of moral vocabulary:

I have dealt more at length with ... 'equal right' and 'fair distribution' ... in order to show what a crime it is to attempt, on the one hand, to force on our Party again, as dogmas, ideas which in a certain period had some meaning but have now become obsolete verbal rubbish, while again perverting, on the other, the realistic outlook, which it cost so much effort to instil into the Party but which has now taken root in it, by means of ideological nonsense about right (*Recht*) and other trash so common among the democrats and French socialists. (Marx 1875: 25.)

And in his letter to Sorge of 19 October 1877, Marx complained about those who wanted to replace socialism's 'materialistic basis (which demands serious objective study from anyone who tries to use it) by modern mythology with its goddesses of Justice, Liberty, Equality and Fraternity' (Marx 1877: 376).

Morality had, moreover, at best a purely derivative role to play in the analysis of capitalist relations and the explanation of social change. In his *Ethnological Notebooks*, Marx annotates a passage in which Henry Sumner Maine writes of 'the vast mass of forces, which we may call for shortness *moral*' which 'perpetually shapes, limits or forbids the actual direction of the forces of society by its Sovereign'. Marx comments that

this 'moral' shows how little Maine understands of the matter; so far as these influences (*economical* before everything else) possess [a] moral modus of existence, this is always derived, secondary modus and never the prius. (Marx 1880–2: 329, S.L.)

Finally, a passage in *The Civil War in France* sums up Marx's continuing denial of the relevance of utopias and ideals to working-class action. The working class, he wrote,

have no ready-made utopias to introduce *par decret du peuple*. They know that in order to work out their own emancipation, and along with it that higher form to which present society is irresistibly tending by its own economical agencies, they will have to pass through long struggles, through a series of historic processes, transforming circumstances and men. They have no ideals to realise, but to set free elements of the new society with which old collapsing bourgeois society itself is pregnant. (Marx 1871a: 523.)

On the other hand, as this last quotation shows, this new society was clearly seen as a 'higher form' of society, and it was plainly the end to which working-class struggle was directed. Interestingly, in a draft of the very same work, *The Civil War in France*, Marx observed that the Utopian Socialists were 'clearly describing the goal of the social movement, the super-session of the wage system with all its economical conditions of class rule'; but they

tried to compensate for the historical conditions of the movement by phantastic pictures and plans of a new society in whose propaganda they saw the true means of salvation. From the moment the workingmen class movement became real, the phantastic utopias evanesced, not because the working class had given up the end aimed at by these Utopists, but because they had found the real means to realise them, but in their place came a real insight into the historic conditions of the movement and a more and more gathering force of the military organisation of the working class. But the last 2 ends of the movement proclaimed by the Utopians are the last ends proclaimed by the Paris Revolution and by the International. Only the means are different and the real conditions of the movement are no longer clouded in utopian fables. (Marx 1871b: 67.)

What, then, was this 'higher form' of society, the means to whose realization was now transparent? Marx characterized it, in the Tenth Thesis on Feuerbach, as '*human* society, or associated humanity' (Marx 1845: 8). As is well known, Marx in turn divided that higher form into two phases, a lower and a higher, which Lenin christened socialism and communism respectively. The former abolishes exploitation but not yet exchange (between the individual and the state); the latter—full communism— represents Marx's ultimate ideal form of social life. Of course, Marx's descriptions of post-capitalist society are extremely thin and scattered throughout his writings. It appears from an 1851 outline of what was to become *Capital* that he intended to present his views about communism in a systematic manner in a final volume. As it was, he never wrote what Engels in a letter to him described as 'the famous "positive", what you "really" want' (cited Ollman 1977: 49).

What is clear is that the ideal society to which Marx expect- antly looked forward would be one in which, under conditions of abundance, human beings can achieve self-realization in a new, transparent form of social unity, in which nature, both physical and social, comes under their control. What is wealth, he asks, 'other than the universality of individual needs, capacities, pleasures, productive forces etc., created through universal exchange? The full development of human mastery over the forces of nature, those of so-called nature as well as humanity's own nature?' (Marx 1857–8: 488). In *Capital* he describes production in such a society as consisting in

socialized man, the associated producers, rationally regulating their interchange with Nature, bringing it under their common control, instead of being ruled by it as by the blind forces of Nature; and achieving this with the least expenditure of energy and under conditions most favourable to, and worthy of, human nature. (Marx 1861–79: iii. 800.)

Beyond this sphere of production, which 'still remains a realm of necessity', there 'begins that development of human energy which is an end in itself, the true realm of freedom, which, however, can blossom forth only with this realm of necessity as its basis' (ibid.) This will enable 'the development of the capacities of the *human* species' (albeit 'at the cost of the majority of human individuals and even classes'), the '*development of the richness of human nature as an end in itself*'

(Marx 1862–3: ii. 117–18). Such a society is variously described by Marx in terms of true freedom and 'real liberation' (Marx and Engels 1845–6: 38), as the overcoming of alienation and the realization of the human essence or human nature, and in utilitarian terms of welfare and happiness (see Kamenka 1969). Above all, however, he seems to have linked three ideas together: (1) the self-realization of individuals, the full development of their essentially human powers, of the 'rich individuality which is as all-sided in its production as in its comsumption, and whose labour therefore no longer appears as labour' (Marx 1857–8: 325); (2) the establishing of new, rational, and harmonious social relations which involves the abolition of 'the contradiction between the interest of the separate individual or the individual family and the interest of all individuals who have intercourse with one another', which in turn involves abolishing the division of labour and private property (Marx and Engels 1845-6: 46); and (3) a prior history of 'total alienation' in which the productive forces are developed by means of a 'sacrifice of the human end-in-itself to an entirely external end' (Marx 1857–8: 487–8), in which 'man's own deed becomes an alien power opposed to him which enslaves him instead of being controlled by him'. (Marx and Engels 1845–6: 47) The linkage of these ideas is well brought out in the following passage from the *Grundrisse*:

Universally developed individuals, whose social relations, as their own communal (*gemeinschaftlich*) relations, are hence also subordinated to their own communal control, are no product of nature but of history. The degree and the universality of the development of wealth where *this* individuality becomes possible supposes production on the basis of exchange values as a prior condition, whose universality produces not only the alienation of the individual from himself and from others, but also the universality and the comprehensiveness of his relations and capacities. (Marx 1857–8: 162.)

Marx's projected future coincides with this idea of a higher form of 'human society' to which he saw humanity as imminently progressing. He made the claim, in a letter to Ruge, that 'we do not anticipate the world dogmatically, but that we first try to discover the new world from a criticism of the old one' (Marx 1843c: 30). But that claim, and his belief that the new world was latent in the womb of the old, does not make his ideal and his criticism any the less evaluative and, in a sense to be explained, moral. And indeed his whole life's work is full of critical judgements that only make sense against the background of this ideal of transparent social unity and individual self-realization.

It lies behind innumerable such judgements from the very beginning. So, in 1843, he wrote that 'the criticism of religion ends with the teaching that *man is the highest being for man*, hence with the *categorical imperative to overthrow all relations* in which man is a debased, enslaved, foresaken and despicable being' (Marx 1844a: 182), and remarked that 'the essential sentiment' of criticism is '*indignation*' and its 'essential activity . . . *denunciation*' (ibid.: 177). It lay behind his description of the working class as 'in its abasement, the *indignation* at that abasement, an indignation to which it is necessarily driven by the contradiction between its human *nature* and its condition of life, which is the outright, resolute and comprehensive negation of that nature' (Marx and Engels 1845: 36); and of capital as 'dead labour, that, vampire-like, only lives by sucking living labour, and lives the more, the more labour it sucks' (Marx 1867: 233). Hence for example all the passages about the stunting effects of the division of labour under capitalism on the labourer, enabling him to achieve 'only a one-sided, crippled development' (Marx and Engels 1845–6: 262), or the way the capitalist mode of production in producing surplus value confronts 'the labourer as powers of capital rendered independent, and standing in direct opposition therefore to the labourer's own development' (Marx 1861–79: iii. 858–9). Hence his talk of English capitalism's 'shameful squandering of human labour power for the most despicable purposes' (Marx 1867: 394). Hence all the passages in *Capital* about 'naked self-interest and callous cash payment', 'oppression', 'degradation of personal dignity', 'accumulation of misery', 'physical and mental degradation', 'shameless, direct and brutal exploitation', the 'modern slavery of capital', 'subjugation', the 'horrors' (surpassing those in Dante's *Inferno* (Marx 1867: 246)) and 'torture' and 'brutality' of overwork, the 'murderous' search for economy in the production process, capital's 'laying waste and squandering' of labour power and 'the natural force of human beings', and capitalism as a system of production 'altogether too prodigal with its human material' and exacting 'ceaseless human sacrifices' (see especially Marx 1867: chs. IV, VII, and XXIII; and

Stojanovic 1973: 137 ff. and Geras 1983: 84). Hence his view that under capitalism 'the devaluation of the world of men is in direct proportion to the *increasing value* of the world of things' (Marx 1844c: 271–2).

How, indeed, can one fail to see the moral force of all these passages, and of those in which Marx speaks of machinery 'transforming the workman, from his very childhood, into a part of a detail-machine', a 'lifeless mechanism independent of the workman, who becomes its mere living appendage' so that 'factory work exhausts the nervous system to the uttermost [and] does away with the many-sided play of the muscles, and confiscates evey atom of freedom, both in bodily and intellectual activity' (Marx 1867: 422–3). Under the capitalist mode of production, according to Marx, 'the labourer exists to satisfy the needs of self-expansion of existing values, instead of, on the contrary, material wealth existing to satisfy the needs of development on the part of the labourer' (ibid.: 621). Consider finally the following passage from *Capital* about the production of relative surplus value within the capitalist system:

all methods for raising the social productiveness of labour are brought about at the cost of the individual labourer; all means for the development of production transform themselves into means of domination over, and exploitation of, the producers; they mutilate the labourer into the fragment of a man, degrade him to the level of an appendage of a machine, destroy every remnant of charm in his work and turn it into a hated toil, they estrange from him the intellectual potentialities of the labour process in the same proportion as science is incorporated in it as an independent power; they distort the conditions under which he works, subject him during the labour process to a despotism the more hateful for its meanness; they transform his life-time into working-time, and drag his wife and child beneath the wheels of the Juggernaut of capital. . . . Accumulation of wealth at one pole is . . . at the same time accumulation of misery, agony of toil, slavery, ignorance, brutality, mental degradation at the opposite pole, i.e. on the side of the class that produces its own product in the form of capital. (Ibid.: 645.)

As for Engels, he wrote this in *Anti-Dühring*:

We therefore reject every attempt to impose on us any moral dogma whatsoever as an eternal, ultimate and for ever immutable ethical law on the pretext that the moral world, too, has its permanent principles which stand above history and the differences between nations. We maintain on the contrary that all moral theories have been hitherto the product, in the last analysis, of the economic conditions of society obtaining at the time. And as society has hitherto moved in class antagonisms, morality has always been class morality; it has either justified the domination and the interests of the ruling class, or, ever since the oppressed class became powerful enough, it has represented its indignation against this domination and the future interests of the oppressed. (Engels 1877–8: 131.)

Of equality, Engels wrote that 'the idea of equality, both in its bourgeois and its proletarian form, is . . . itself a historical product, the creation of which required definite historical conditions that in turn themselves presuppose a long previous history. It is therefore anything but an eternal truth.' As for justice, he deplored 'idealistic phraseology about justice' (Engels 1890: 496), and remarked that:

According to the laws of bourgeois economics, the greatest part of the product does *not* belong to the workers who have produced it. If we now say: that is unjust, that ought not to be so, then that has nothing immediately to do with economics. We are merely saying that this economic fact is in contradiction to our sense of morality. Marx, therefore, never based his communist demands upon this, but upon the inevitable collapse of the capitalist mode of production which is daily taking place before our eyes to an ever greater degree. (Engels 1884: 18.)

Justice, Engels wrote, attacking Proudhon, is 'but the ideologised, glorified expression of the existing economic relations, now from their conservative, and at other times from their revolutionary angle':

While in everyday life, in view of the simplicity of the relations discussed, expressions like right, wrong and sense of right are accepted without misunderstanding even with reference to social matters, they create, as we have seen, the same hopeless confusion in any scientific investigation of economic relations as would be created, for instance, in modern chemistry if the terminology of the phlogiston theory were to be retained. The confusion becomes still worse if one, like Proudhon, believes in this social phlogiston, 'justice'. (Engels 1872–3: 624–5.)

Communism, Engels wrote,

now no longer meant the concoction, by means of the imagination, of an ideal society as perfect as possible, but insight into the nature, the conditions, and the consequent general aims of the struggle waged by the proletariat. (Engels 1885: 345.)

And yet, on the other hand, Engels also wrote in *Anti-Dühring* of 'the proletarian morality of the future', as 'containing the maximum elements producing permanence'; and he also argued that the claim that

there has on the whole been progress in morality, as in all other branches of human knowledge, no one will doubt. But we have not yet passed beyond class morality. A really human morality which stands above class antagonisms and above any recollection of them becomes possible only at a stage of society which has not only overcome class antagonisms but has even forgotten them in practical life. (Engels 1877–8: 130, 132.)

In earlier speeches and writings Engels had certainly been prepared to sketch the outlines of such a society (Engels 1845a), and, as we have seen, his writings are as replete as those of Marx with sharp and biting morally-based critiques of class societies.

The only sustained treatment of these questions in the classical marxist canon is Karl Kautsky's *Ethics and the Materialist Conception of History*. It is important to note that Kautsky was writing in the context of the rise of neo-kantianism, especially in Germany and Austria. To the German kantian-influenced marxists and sympathizers with marxism, the kantian ethic could provide the foundation that would show that socialism, even if inevitable, should be accepted as desirable and worth striving for. The Marburg School in particular, beginning with F. A. Lange, generally believed that socialism, once provided with such a foundation, represented the practical, political expression of kantian ethical principles (see Keck 1975). The neo-kantians' opponents—such as Mehring who saw them as trying to 'throttle the entire socialist movement' (*Die neue Zeit*, II (1900), 1 cited in ibid.; see Sandkuehler and de la Vega 1970: 41, 42), Rosa Luxemburg, and Plekhanov—thought no such foundation was necessary; but, as Staudinger saw, explicability, even when conjoined with inevitability, does not equal desirability: a rotten apple can only be the way it is, but it is rotten for all that (cited in Kolakowski 1978: ii. 251). Vorländer stated the position clearly:

Precisely because marxism . . . as a social historical theory must necessarily exclude the ethical standpoint, it is, in our opinion, all the more essential for the foundation and justification of socialism to take into account this complementary standpoint, which was and is an integral element in the history of socialist ideas and even more in socialist practice. (Vorländer 1921: 431.)

Even when 'in ostensible opposition to ethics and idealism, a deep ethical intuition is latent. . . . Socialism can divorce itself from ethics neither historically nor logically, neither theoretically nor factually' (Vorländer 1904: 23; in support of which Vorländer cited Marx's frequent use of value-laden terminology). It is worth noting that the German neo-kantians were not by intention revisionists (though Bernstein took up some of their slogans), for they were trying to plug what they saw as a gap in marxism. They saw kantian ethics as a natural completion of marxism: it furnished the formal definition of the conditions which any moral precept must fulfil, while marxism set out which concrete actions would lead to the aims it shared with kantianism, namely universal brotherhood and solidarity, based on the recognition of the irreducible value of every human individual. But did that recognition preclude the use of revolutionary tactics and, in particular, of violence? On the whole, the Marburg neo-kantians thought that it did.

The German neo-kantians—Hermann Cohen, Natorp, Stammler, Staudinger, and Vorländer—sought, in short, to supplement Marx with Kant, whose practical philosophy, they thought, could provide the ethical justification for the pursuit of the socialist goal. The Austro-marxists—especially Max Adler and Otto Bauer—were also influenced by Kant, but firmly rejected this position, arguing that socialism did not stand in need of such justification. Adler maintained that 'according to the Marxist conception socialism does not come about because it is ethically justified, but because it is causally produced' (Adler 1925: 64). This 'causal product of social life' was, however, 'at the same time ethically justified'. This was because 'socialized man . . . is finally driven by formal-teleological causality to realise what he considers to be morally justified' (ibid.: 64). Bauer too opposed 'the attempt of some revisionists to import Kant's principle of practical reason into the justification of socialism' (Bauer 1906: 83). Marx, he thought, had scientifically demonstrated that, in capitalist society, the proletariat was bound to *want* socialism as the only possibility of escaping exploitation; that it *can* attain its goal because the concentration of property has made possible the appropriation of the instruments of labour as social property; that the working class *will* attain its goal, because it becomes the overwhelming majority of the population. (Ibid.: 80–1.)

But what of the 'moral question' faced by 'the still hesitant and undecided individual' yet to 'decide which of the contradictory classes he will join' (ibid.: 81)? Bauer's answer was in effect that Kant's ethics were implicit in the very struggle of the proletariat, and did not need to be made explicit:

If we only make him realise that the proletariat necessarily fights against exploitation, that with the development of capitalist society the fight against exploitation necessarily leads to a struggle for the socialist mode of production, we give him the material for a decision, and the decision itself will then be the correct judgement on the rightness of his will. Without ever having heard of Kant's categorical imperative, he will immediately judge the maxims flowing from the class interest of the proletariat differently from those of the classes defending their property. He will judge as immoral the maxims of those who have to defend a social order which can only exist if it represses the overwhelming majority of the members of society, either by force, or by deceiving them about their real interests by means of a false ideology. (Ibid.: 80–1.)

Kantian ethics was, therefore, not needed as a justification, but it did serve as

the final bastion to which we can retreat whenever ethical scepticism obstructs the naïve moral judgement of class maxims discovered by science. . . . If anyone misled by ethical scepticism thinks that there is no criterion of choice for him, because he knows the necessary will of all classes, we recall to his attention the formal law-governed character of his will, we provide the criterion which enables him to distinguish the will of the working class from that of the bourgeoisie in terms of its value, and so guide him into the camp of the fighting proletariat. (Ibid.: 82–3.)

Bauer differed from Kautsky in his 'conviction that we cannot renounce Kant's critique of reason; time and again it is able to protect us from the storm of scepticism unleashed by the enemies of the working class' (ibid.: 84).

Kautsky was unimpressed by Bauer's attempted rejoinder to the ethical sceptic. If, he wrote, instead of hastily rushing to Kant, the sceptic aimed to be correctly informed about marxist ethics, he would recognise not only that no ethic is absolute and that moral rules can change, but also that ethical rules are necessary for particular times, societies and classes; that ethics are not a matter of convention, nor something which the individual chooses at will, but are determined by powers which are stronger than the individual, which stand over him. How can scepticism arise out of the recognition of necessity? (Kautsky 1906b: 49–50.)

From kantian circles, Kautsky plausibly added,

we never get a clear answer to the question of which ethic is the more valuable, the bourgeois or the proletarian; among them, we find the most diverse views about this question, but conservatives and liberals are far more strongly represented than socialists. Therefore the categorical imperative does not necessarily lead our sceptic towards socialism. . . . Even for the Kantian, it is his economic conception and insight which helps him decide which side to take, not the categorical imperative. (Ibid.: 50–1.)

Yet, as late as 1898, Kautsky had written to Plekhanov that he was 'troubled by neo-kantianism. I believe . . . that the economic and historical standpoint of Marx and Engels is compatible with neo-kantianism'—though he also admitted that 'philosophy was never my strong suit' (cited in Steinberg 1967: 99). But he soon recanted and came to his settled view that marxism was sufficient unto itself and needed no ethical expansion, and that the neo-kantians offered an anachronistic ethic of reconciliation. In arguing thus, he echoed Marx's and Engels's criticisms of Proudhon and Lasalle, speaking with scorn of 'Ethical Socialism' as 'endeavours . . . in our ranks to modify the class antagonisms, and to meet at least a section of the Bourgeoisie half way', the 'historical and social tendency' of the kantian ethic being that of 'toning down, of reconciling the antagonisms, not of overcoming them through struggle' (Kautsky 1906a: 69). For Kautsky, moral tenets 'arise from social needs', 'all morality is relative', what is 'specifically human in morality, the moral codes, is subject to continual change': 'the moral rules alter with the society, yet not uninterruptedly and not in the same fashion and degree as the social needs', indeed 'it is with the principles of morality as with the rest of the complicated social superstructure which arises upon the means of production. It can break away from its foundation and lead an independent life for a time' (ibid.: 178, 192, 184). It was, he writes, 'the materialist conception of history which has first completely deposed the moral ideal as the directing factor of social evolution, and has taught us to deduce our social aims solely from the knowledge of the material foundations' (ibid.: 201). Even with Marx, Kautsky says,

occasionally in his scientific research there breaks through the influence of a moral ideal. But he always endeavours, and rightly, to banish it where he can. Because the moral ideal becomes a source of error in science, when it presumes to point out to it its aims. Science has only to do with the recognition of the necessary. It can certainly arrive at prescribing an *ought* but this can arise only as a result of insight into what is necessary. It must decline to discover an *ought* which is not recognisable as a necessity founded in the world of phenomena. (Ibid.: 202–3, S.L.)

As Kolakowski has observed, Kautsky

agreed with the neo-kantians that marxism proved the historical necessity of socialism, and this in his view was all that required to be shown. The working class was bound to develop a consciousness that would regard socialism as an ideal, but this attitude of mind was itself no more than the consequence of a social process. The question why a person should regard as desirable what he believes to be inevitable is ignored by Kautsky, who gives no reason for not answering it. (Kolakowski 1978: ii. 39.)

On the other hand, as Vorländer noticed, Kautsky admitted that norms and goals were integral to human experience. 'The conscious aim of the class struggle in Scientific Socialism', he wrote, 'has been transformed from a moral into an economic aim', but then he added upliftingly that 'it loses none of its greatness' (Kautsky 1906: 204). Kautsky seems to have supposed that because what had previously been seen as a moral ideal could now be recognized as a necessary result of economic development, it thereby ceased to be a moral ideal. Of that ideal—'the abolition of class . . . of all social distinctions and antagonisms which arise from private property in the means of production and from the exclusive chaining down of the mass of the people in the function of production . . . of rich and poor, exploiters and exploited, ignorant and wise . . . of the subjection of women by men [and] of national antagonisms'—he asks:

'Where is there a moral ideal which opens up such splendid vistas?' And indeed, he maintained that with the expansion of capitalism, there had formed the basis for a 'general human morality' (ibid.: 204–6, 159). This was formed by

the development of the productive forces of man, by the extension of the social division of labour, the perfection of the means of intercourse. This new morality is, however, even today far from being a morality of all men even in the economically progressive countries. It is in essence even today the morality of the class-conscious proletariat, that part of the proletariat which in its feeling and thinking has emancipated itself from the rest of the people and has formed its own morality in opposition to the bourgeoisie. (Ibid.: 159–60.)

It is, he adds, 'capital which creates the material foundation for a general human morality, but it only creates the foundation by treading this morality continually under its feet' (ibid.: 160). On this basis, the proletariat will 'create a form of society in which the equality of man before the moral law will become—instead of a mere pious wish—reality' (ibid.: 160–1).

Plekhanov, like Kautsky, defended orthodoxy against revisionism and neo-kantianism which he described as 'not a fighting philosophy' but 'a philosophy of persons who, when all is said and done, stop half way: it is a philosophy of compromise.' Resisting Stammler's thought that social relations 'are the outcome of actions undertaken to attain a foreseen end', he saw no need to specify the ends or to supply moral grounds for joining a movement certain to succeed (Plekhanov 1908: 94, 90 ff.). He seems simply to have supposed that since human purposes are determined by productive forces and social conditions, the provision of such grounds is rendered unnecessary. As he wrote,

When a class longing for emancipation brings about a social revolution, it acts in a way which is more or less appropriate to the desired end; and in any case, its activity is the cause of that revolution. But the activity, together with all the aspirations which have brought it about, is itself the effect of ecomonic evolution, and therefore, is itself determined by necessity. (Ibid.: 92–3.)

Nor did Plekhanov think of moral principles as guiding or setting limits to such activity. As he said in a famous and prophetic speech to the 1903 Congress,

Every democratic principle must be considered not by itself, abstractly, but in relation to what may be called the fundamental principle of democracy, namely *salus populi suprema lex*. Translated into the language of the revolutionist, this means that the success of the revolution is the highest law. And if the success of the revolution demanded a temporary limitation of the working of this or that democratic principle, then it would be criminal to refrain from such a limitation. As for my own personal opinion, I will say that even the principle of universal suffrage must be considered from the point of view of what I have designated the fundamental principle of democracy. It is hypothetically possible that we, the Social Democrats, might speak out against universal suffrage. The bourgeoisie of the Italian Republics once deprived persons belonging to the nobility of political rights. The revolutionary proletariat might limit the political rights of the higher classes just as the higher classes once limited their political rights. One can judge of the suitability of such measures only on the basis of the rule: *salus revolutiae suprema lex*.

And we must take the same position on the question of the duration of parliaments. If in a burst of revolutionary enthusiasm the people chose a very fine parliament—a kind of *chambre introuvable*—then we would be bound to make it a *long parliament*; and if the elections turned out unsuccessfully, then we would have to try to disperse it not in two years but if possible in two weeks. (Plekhanov 1903: 106.)

On the other hand, Plekhanov spoke of the proletariat as representing 'the supreme interests of mankind' (Plekhanov 1908: 96), and of Marx's 'calm and virile faith that the final aim will in due course be achieved', that final aim being

an end to the exploitation of man by man, and thus to the division of society into a class of exploiters and exploited [which] will make civil wars, not only useless, but impossible. Thenceforward, mankind will advance by the sole 'power of truth' and will no longer have occasion for the argument of the mailed fist. (Ibid.: 109.)

Plekhanov even wrote that 'the more energetically a person fights for the realisation of his social ideals, the greater his self-sacrifice in this struggle, the higher he steps on the rungs of the ladder of moral perfection' (cited Heller 1982: 364). Yet, given Plekhanov's views about morality and the irrelevance of moral considerations, it is altogether mysterious why anyone should undertake such self-sacrifice. Only if proletarians' sense of dignity and aspirations for a better world provide grounds for action which override their immediate and private interests, does their self-sacrifice in the class struggle make any sense—a truth Marx himself recognized when he wrote that 'the proletariat, which will not permit itself to be treated as rabble, needs its courage, its self-confidence, its pride and its sense of independence more than its bread' (Marx 1847a: 231).

Lenin likewise fiercely combated Ethical Socialism and what he called the subjective and moralistic viewpoint of populist writers, which he contrasted with marxism as a scientific determinist doctrine which does not ask questions about what ought to be but considers all processes including the phenomena of consciousness as natural events determined by the relations of production. Lenin indeed describes the proletariat as 'the intellectual and moral motive force and the physical executor' of 'the inevitable advent of socialism', but states that for Marx that inevitability is deduced 'wholly and exclusively from the economic law of development of contemporary society' (Lenin 1915: 71). Marx, he wrote,

treats the social movement as a process of natural history, governed by laws not only independent of human will, consciousness and intentions, but, on the contrary, determining the will, consciousness and intentions of men. . . . if the conscious element plays so subordinate a part in the history of civilisation, it is self-evident that a critique whose subject is civilisation can least of all take as its basis any form of, or any result of, consciousness. (Lenin 1893: 166.)

Everybody knows, Lenin wrote, that 'scientific socialism never painted any prospects for the future as such: it confined itself to analysing the present bourgeois regime, to studying the trends of development of the capitalist social organisation, and that is all' (ibid.: 184). There was, he remarked in *The State and Revolution*,

no trace of an attempt on Marx's part to make up a utopia, to indulge in idle guess-work about what cannot be known. Marx treated the question of communism in the same way as a naturalist would treat the question of the development of, say, a new biological variety, once he knew that it had originated in such and such a way and was changing in such and such a definite direction. (Lenin 1917a: 458.)

Lenin agreed with Sombart that there was in marxism 'not a grain of ethics from beginning to end' since, 'theoretically, it subordinates the "ethical standpoint" to the "principle of causality": in the practice, it reduces to the class struggle'

(Lenin 1894: 421). And there are many passages in Lenin's writings where he takes an aggressively instrumentalist view of morals, arguing, for example, that moral rules about labour discipline could properly be ignored by the proletariat under bourgeois rule, though their practical application by the Soviet state was a condition for the final victory of Socialism. And in 1920 he told the Komsomol Congress:

> We say that our morality is entirely subordinated to the interests of the proletariat's class struggle. . . . Morality is what serves to destroy the old exploiting society and to unite all the working people around the proletariat, which is building up a new, a communist society. . . . To a communist all morality lies in this united discipline and conscious mass struggle against the exploiters. We do not believe in an eternal morality, and we expose the falseness of all the fables about morality. (Lenin 1920: 291–4.)

Yet on the other hand, Lenin's writings are full of passionate moral denunciations of the ills of capitalism, as when he wrote in 1917 of 'these survivals of accursed capitalist society, these dregs of humanity, these hopelessly decayed and atrophied limbs, this contagion, this plague, this ulcer that socialism has inherited from capitalism' (Lenin 1917b: 410). In *The State and Revolution*, Lenin clearly suggests that Marx's higher phase of communism will consist in a more *just* distribution of consumer goods, and writes of the lower phase as 'by no means our ideal, or our ultimate goal. It is only a necessary *step* for cleansing society of all the infamies and abominations of capitalist exploitation *and for further* progress' (Lenin 1917a: 466, 474). In the process of that development, 'the working man can reveal his talents, unbend his back a little, rise to his full height, and feel that he is a human being' (Lenin 1917b: 407); while, in more utilitarian vein, socialism alone 'will make possible the wide expansion of social production and distribution along scientific lines and their actual subordination to the aim of easing the lives of working people and of improving their welfare as much as possible' (Lenin 1918: 411). The proletariat, he wrote in 1919, were 'not only overthrowing the exploiters and suppressing their resistance, but are building a new and higher social bond, a social discipline, the discipline of class-conscious and united working people' (Lenin 1919: 423); and in 1920 he wrote of communist labour as 'labour performed because it has become a habit to work for the common good, and because of a conscious realisation (that has become a habit) of the necessity of working for the common good—labour as the requirement of a healthy organism'. It will, he proclaimed, 'take many years, decades to create a new labour discipline, new forms of social ties between people, and new forms and methods of drawing people into labour. It is a most gratifying and noble work' (Lenin 1920: 518). Materialism, for Lenin, 'includes partisanship, so to speak, and enjoins the open and direct adoption of the standpoint of a definite social group in any assessment of events' (Lenin 1894: 401). Was that standpoint not self-evidently moral? As Trotsky observed,

the 'amoralism' of Lenin, that is, his rejection of supra-class morals, did not hinder him from remaining faithful to one and the same ideal throughout his life; from devoting his whole being to the cause of the oppressed; from displaying the highest conscientiousness in the sphere of ideas and the highest fearlessness in the sphere of action; from maintaining an attitude untainted by the least superiority to an 'ordinary' worker, to a defenceless woman, to a child. Does it not seem that 'amoralism' in the given case is only a pseudonym for higher human morality? (Trotsky 1938: 34.)

As for Trotsky's own views, he expressed these often, fiercely and honestly, and nowhere more so than in his *Terrorism and Communism* (to be discussed in Chapter 6), where he wrote that:

As for us, we were never concerned with the Kantian-priestly and vegetarian-Quaker prattle about 'sacredness of human life'. We were revolutionaries in opposition, and have remained revolutionaries in power. To make the individual sacred we must destroy the social order which crucifies him. And this problem can only be solved by blood and iron. (Trotsky 1920: 82.)

But it is his celebrated debate with John Dewey, initiated by his remarkable pamphlet, *Their Morals and Ours*, that reveals his general views about morality most clearly. Here he reveals the same pattern of thought and the same resulting paradox that we have been considering. Trotsky enjoins his readers to agree that 'morality is a function of the class struggle' (Trotsky 1938: 18) and 'a product of social development; that there is nothing immutable about it; that it serves social interests; that these interests are contradictory; that morality more than any other form of ideology has a class character' (ibid.: 15). Though a few 'elementary moral precepts exist, worked out in the development of mankind as a whole and indispensable for the existence of every collective body', their influence is 'extremely limited and unstable' (ibid.); 'Norms "obligatory upon all" become the less forceful the sharper the character assumed by the class struggle.' Indeed, the 'highest form of the class struggle is civil war which explodes into mid-air all moral ties between the hostile classes' (ibid.: 16, 15–16). For 'The norms of "obligatory" morality are in reality filled with class, that is, antagonistic content. . . . The solidarity of workers, especially of strikers and barricade fighters, is incomparably more "categoric" than human solidarity in general' (ibid.: 16). 'Moral evaluations', for Trotsky, 'together with those political, flow from the inner needs of struggle' (ibid.: 38).

On the other hand, Trotsky also writes, 'to a revolutionary marxist' the party embodies 'the very highest tasks and aims of mankind' (ibid.: 34); and, discussing the relation between means and ends, he states that

A means can be justified only by its end. But the end in its turn needs to be justified. From the Marxist point of view, which expresses the historical interests of the proletariat, the end is justified if it leads to increasing the power of man over nature and to the abolition of the power of man over man. . . . That is permissible . . . which *really* leads to the liberation of mankind. Since this end can be achieved only through revolution, the liberating morality of the proletariat of necessity is endowed with a revolutionary character. It irreconcilably counteracts not only religious dogma but all kinds of idealistic fetishes, these philosophic gendarmes of the ruling class.

But then he adds a sentence on which, as we shall see, Dewey focused his critical attention: 'It deduces a rule of conduct from the laws of development of society, thus primarily from the class struggle, this law of all laws' (ibid.: 36–7). In short, like Marx, Engels, Kautsky, and Lenin, Trotsky was committed on the one hand to the moral condemnation of capitalist evils and the advocacy and pursuit of socialist ends, and indeed to the justification of these ends in terms of a 'liberating morality'; and on the other, to the dismissal of all moral talk as dangerous ideological illusion, rendered anachronistic by the discovery of scientific laws of economic development.

I shall not go any further in documenting this paradox. Marxism is a bibliocentric creed, and so the presence of this paradox in the writings of these classical marxists is enough to guarantee its continuing impact among their many and varied descendants. Obviously, it is most apparent in all those versions of orthodox and 'scientific' marxism that are propagated in Institutes of Marxism–Leninism and taught in party schools in the Soviet bloc and China, as well as in Europe and the Third World. But it has been by no means absent from 'neo-marxism' or the various varieties of so-called 'Western marxism'. Of course, in neo-marxist thought it takes on more nuanced forms, along with increasing sophistication about the nature of economic determinism, ideology, and science. But, with some notable exceptions—notably Gramsci, Walter Benjamin, and Ernst Bloch, and critical and humanist marxists of the postwar period, especially in Yugoslavia—marxism has remained, in its distinctive and curious way, both anti-moral and moral.

In large part, this fact betokens marxism's historic unreflectiveness about moral matters. Ever since Marx, it has not brought its official critical position *vis-à-vis* morality and its actual moral commitments into relation with one another. On the matter of moral values, Edward Thompson has observed, 'the silence of Marx, and of most Marxisms is so loud as to be deafening.' Noting, as we have done, that 'Marx, in his wrath and compassion, was a moralist in every stroke of his pen', Thompson argues that Marx's and Engels's battle against the triumphant moralism of Victorian capitalism led them to this neglect and silence. Moreover,

This silence was transmitted to the subsequent Marxist tradition in the form of a repression. This repression, in its turn, made it more easy for the major tradition to turn its back upon Morris (and many other voices) and to capitulate to an economism which, in fact, simply took over a bourgeois utilitarian notion of 'need'; and, as a necessary complement to this, to foster a paltry philistinism towards the arts. It was only necessary for Marxist Science to enter the kingdom of Socialism, and all else would be added thereunto. And Marxism-Leninism-Stalinism did. And we know with what results. (Thompson 1978: 363, 364.)

In his reply to Thompson, Perry Anderson agrees that 'Marx and Engels left no Ethics, and that the resultant gap was never made good in the Marxism which ensued after their deaths—to the danger of historical materalism as a theory and of the socialist movement as a practice.' He stresses, as both Thompson and I have done, that 'Neither Marx nor Engels . . . was in any way reluctant to express solid moral judgments. They did not, however, systematise these into a separate discourse', and he offers a reason why 'the founders of historical materialism were so chary of ethical discussions of socialism': such discussions tend

to become *substitutes* for explanatory accounts of history. Agressively claiming to reinstate 'moralism' as an integral part of any culture of the Left, Thompson has forgotten the distinction which the term itself is designed to indicate in ordinary usage. Moral consciousness is certainly indispensable to the very idea of socialism: Engels himself emphasised that 'a really human morality' would be one of the hallmarks of communism, the finest product of its conquest of the age-old social divisions and antagonisms rooted in scarcity. Moralism, on the other hand, denotes the vain intrusion of moral judgments in lieu of causal understanding—typically, in everyday life and in political evaluations alike, leading to an 'inflation' of ethical terms themselves into a false rhetoric, which lacks the exacting sense of material care and measure that is inseparable from true moral awareness. This process is all too evident and familiar in contemporary politics outside the socialist movement, and against it. Solzhenitsyn since his exile is a signal example. Its end result is to devalue the writ of moral judgment altogether. (Anderson 1980: 97–8, 86.)

There is undoubtely something in this. It offers perhaps the best reason there is for marxism's avoidance of 'morality': moral discourse, it is suggested, all too easily smothers the explanatory enterprise, whose careful and strenuous pursuit is integral to 'true moral awareness'. But we still need to interpret that avoidance and assess its consequences. Anderson's suggestion leaves untouched the question we are here seeking to address: is there a consistent view of morality that resolves the paradox we have been discussing and makes sense both of marxism's general rejection of morality and its actual (if unsystematic and largely unacknowledged) moral stance? It is to that question that the next chapter is addressed.

3 The Paradox Resolved

The moment anyone started to talk to Marx about morality, he would roar with laughter.

(Vorländer 1904: 22.)

[Marx's] real mission in life was to contribute, in one way or another, to the overthrow of capitalist society, and of the state institutions which it had brought into being, to contribute to the liberation of the modern proletariat, which *he* was the first to make conscious of its own position and its needs, conscious of the conditions of its emancipation. Fighting was his element. And he fought with a passion, a tenacity and a success such as few could rival.

(Engels 1883: 168.)

The key to resolving the paradox lies, I believe, in drawing a distinction, to be found in Marx, between what I shall call, following Marx, the morality of *Recht* and the morality of *emancipation*.

Marx and Engels scorned 'the faith of individuals in the conceptions of *Recht*', conceptions which 'they ought to get out of their heads' (Marx and Engels 1845–6: 362). 'As far as *Recht* is concerned,' they wrote, 'we with many others have stressed the opposition of communism to *Recht*, both political and private, as also in its most general form as the rights of man' (Marx and Engels 1845–6: 209). This is an accurate statement about all their writings—and indeed about the marxist tradition in general—from 'On the Jewish Question' onwards.

In that work, Marx spoke of 'the so-called *rights of man*' as 'nothing but the rights of a *member of civil society*, i.e. the rights of egoistic man, of man separated from other men and from the community'. The 'right of man to liberty is based not on the association of man with man, but on the separation of man from man. It is the *right* of this separation, the right of the *restricted* individual, withdrawn into himself' (Marx 1843a: 162), its practical application being the right to private property.

This right, the 'right to enjoy one's property and to dispose of it at one's discretion, without regard to other men, independently of society, the right of self interest', Marx saw as forming 'the basis of civil society', making 'every man see in other men, not the *realization* of his own freedom, but the *barrier* to it' (ibid.: 163). In general,

None of the so-called rights of man . . . go beyond egoistic man, beyond man as a member of civil society, that is, an individual withdrawn into himself, into the confines of his private interests and private caprice and separated from the community. In the rights of man, he is far from being conceived as a species being; on the contrary, species-life itself, society, appears as a framework external to the individuals, as a restriction of their original independence. The sole bond holding them together is natural necessity, need and private interest, the preservation of their property and their egoistic selves. (Ibid.: 164.)

The *political community* is 'a mere *means* to maintain these so-called rights of man' (ibid.). 'Human emancipation' contrasts with these rights of man: it will be accomplished only when 'the real, individual man reabsorbs in himself the abstract citizen; and as an individual human being has become a *species-being* in his everyday life, in his particular work, and in his particular situation, only when man has recognized and organized his "*forces propres*" as *social* forces, and consequently no longer separates social power from himself in the shape of *political* power' (ibid.: 168).

'*Recht*', like the French '*droit*' and the Italian '*diritto*', is a term used by continental jurists for which there is no direct English translation. As H. L. A. Hart has observed, these expressions

seem to English jurists to hover uncertainly between law and morals, but they do in fact mark off an area of morality (the morality of law) which has special characteristics. It is occupied by the concepts of justice, fairness, rights and obligation (if this last is not used as it is by many moral philosophers as an obscuring general label to cover every action that morally we ought to do or forbear from doing. (Hart 1955: 177–8.)

'Emancipation' is a term which derives from the Latin '*emancipare*', which in turn derives from '*e + manus + capere*' meaning 'to set free a child or wife from the *patria potestas*' and later of course to set free from slavery, and hence generally from civil disabilities. For Marx human emancipation denoted a setting free from the pre-history of human bondage, culminating in wage-slavery and exploitation, and thus it refers to that ideal of transparent social unity and individual self-realization in which 'the contradiction between the interest of the separate individual or the individual family and the common interest of all individuals who have intercourse with one another has been abolished' (Marx and Engels 1845–6: 46); which I sketched in the previous chapter.

My suggestion, then, is this: that the paradox in marxism's attitude to morality is resolved once we see that it is the morality of *Recht* that it condemns as ideological and anachronistic, and the morality of emancipation that it adopts as its own. Indeed, as I shall argue in this chapter, human emancipation in part precisely consists in emancipation from *Recht*, and the conditions that call it into being.

How did Marx and Engels and their followers conceive of *Recht*? Marx wrote of relations governed by *Recht (Rechtsverhältnisse)* that, 'like forms of state, [they] are to be grasped neither from themselves nor from the so-called general development of the human mind, but rather have their roots in the material conditions of life, the sum total of which Hegel . . . combines under the name of "civil society" whose anatomy is to be sought in political economy' (Marx 1859: 362). For Hegel, 'civil society' meant 'the war of each against all' to be found in the capitalist market-place: it denoted the competitive, egoistic relations of emergent bourgeois society, in which individuals pursued their respective particular interests, treating one another as means to their respective ends, and exercising what Hegel called 'subjective freedom'. Hegel saw certain rights and principles as governing such relations (e.g. private property rights and the principles of contractual justice), and he saw the state as the sphere of citizenship, of 'objective freedom', institutionalizing internally-accepted norms of ethical life and providing the framework within which the mutually destructive forces of civil society could be contained. In this way a rational synthesis of subjective and objective freedom is attained in the modern bourgeois state.

For Marx, on the contrary, both the rights and the principles governing the relations of civil society, and the state itself, were rooted in and means of stabilizing the production relations and thus the class relations of a given social order. The principles of *Recht* were to be understood only in this perspective. As Engels wrote, 'social justice or injustice is decided by the science which deals with the actual facts of production and exchange, the science of political economy' (cited Wood 1972: 15). In short, the principles of *Recht* are to be understood neither (through themselves) as a set of objective norms, a set of independent rational standards by which to assess social relations, nor, following Hegel, as a rational way of ordering such relations, finally uniting subjective with objective freedom, but rather must themselves always in turn be explained as arising, like the social relations they govern and stabilize, out of given material conditions.

This suggests the first marxist reason for opposing *Recht*, namely, that it is inherently ideological. It claims to offer 'objective' principles specifying what is 'just' and 'fair' and defining 'rights' and 'obligations'; it claims that these are universally valid and serve the interests of all the members of society (and perhaps all members of any society); and it claims to be autonomous of particular partisan or sectional interests. But from a marxist point of view all these claims are spurious and illusory. They serve to conceal the real function of principles of *Recht*, which is to protect the social relations of the existing order, a function that is better fulfilled to the extent that the claims are widely accepted as 'objectively' valid. Marxism, in short, purports to unmask the self-understanding of *Recht* by revealing its real functions and the bourgeois interests that lurk in ambush behind it.

It does not, of course, follow from this that all communists should become 'immoralists', violating every bourgeois right and obligation in sight. That would, in any case, be poor tactics. What does follow is that the principles of *Recht* should have for them no rationally compelling force. And it follows from *this* that it makes no sense to criticize capitalism for failing to live up to such principles; for being unjust, violating the rights of workers, etc. (except as a tactical move).

But there is a further and deeper reason for communism's opposition to *Recht*, which can be unearthed if we ask the question: to what problem are the principles of *Recht* a response? To this question jurists and philosophers give different answers, but these answers have in common a view of human life as inherently conflictual, and potentially catastrophically so, thus requiring a framework of authoritative rules, sometimes needing coercive enforcement, that can be rationally justified as serving the interests of all. *Recht* is a response to what one might call the 'conditions of *Recht*', inherent in the human condition, and these may be more or less acute, just as the response will take different forms in different societies.

Consider David Hume's summary account of the conditions of *Recht*: for Hume, "tis only from the selfisheness and confin'd generosity of men, along with the scanty provision nature has made for his wants, that justice derives its origin' (Hume 1739: 495). In his recent book on ethics, John Mackie, citing this statement of Hume's, alongside Protagoras and Hobbes, sought to identify what he calls a 'narrow sense of morality' (which looks very much like *Recht*) as 'a system of a particular sort of constraints on conduct—ones whose central task is to protect the interests of persons other than the agent and which present themselves to an agent as checks on his natural inclinations or spontaneous tendencies to act'. Mackie argues, following Hume, that morality—in the narrow sense thus defined—is needed to solve a basic problem in the human predicament, that 'limited resources and limited sympathies together generate both competition leading to conflict, and an absence of what would be mutually beneficial co-operation' (Mackie 1977: 106, 111). Or consider Kant's celebrated discussion of man's 'unsocial sociality' and of the problem to which the *Rechtstaat* is the solution:

Given a multitude of rational beings who, in a body, require general laws for their own preservation, but each of whom, as an individual, is secretly inclined to exempt himself from this restraint: how are we to order their affairs and establish for them a constitution such that, although their private dispositions may be really antagonistic, they may yet so act as a check upon one another, that, in their public relations, the effect is the same as if they had no such evil sentiments. (Kant 1795: 154.)

Or consider finally John Rawls's account of what I have called the conditions of *Recht* and he calls the 'circumstances of justice': these are 'the normal conditions under which human co-operation is both possible and necessary', and they 'obtain wherever mutually disinterested persons put forward conflicting claims to the division of social advantages under conditions of moderate scarcity' (Rawls 1971: 126, 128). These are 'elementary facts about persons and their place in nature' and, for justice to obtain, 'human freedom is to be regulated by principles chosen in the light of these natural restrictions' (ibid.: 257).

These various suggestions combine to identify three jointly sufficient conditions for the need for justice and rights. (Whether in their imagined absence there would be a need for principles of justice and the recognition of rights is a question largely unaddressed within marxism. We shall raise it in the next chapter). Clearly scarcity, or limits to desired goods,[1] and egoism, or at least the absence of total and unconditional altruism, generate conflicting claims, and thus the need to adjudicate upon which claims are valid, and of these which have priority. More deeply (and this is what Rawls's account implicitly adds to Hume's and Kant's) it is the conflict of interests, resulting from different individuals' (and groups') different and conflicting conceptions of the good, that define those interests, which renders such adjudication, and the protections rights afford, necessary. Hume mistakenly thought that if you increase 'to a sufficient degree the benevolence of men, or the bounty of nature . . . you render justice useless,

---

[1] Of course, scarcity is not a simple notion. Consider the following forms of it: (1) insufficiency of production inputs (e.g. raw materials) relative to production requirements; (2) insufficiency of produced goods relative to consumption requirements; (3) limits upon the possibility of the joint realization of individual goals, resulting from external conditions (e.g. limitations of space or time); and (4) limits upon the possibility of the joint realization of individual goals, resulting from the nature of those goals (e.g. 'positional goods': we cannot all enjoy high status or the quiet solitude of our neighbourhood park). These possibilities bring into view a range of determinants of scarcity, of which nature's niggardliness and men's wants are only two (and these are themselves dependent variables, in turn determined by a range of social, cultural, scientific, and technical factors). Scarcity (2) can exist without scarcity (1): it may result entirely from the existing system of production and distribution. Scarcity (2) can be absent despite limited resources. And scarcity (3) and (4) may result from social, organizational, and cultural factors and exist without scarcity (1) or (2). All these forms of scarcity can generate conflicts of interest. Furthermore, overcoming them all would involve an immense growth in the productive forces of society, changes in social organization, and appropriate preference changes, eliminating non-compatible desires. I shall in the text call this unrealizable state of affairs 'co-operative abundance'.

by supplying its place with much nobler virtues, and more favourable blessings' (Hume 1739: 494–5). But even under conditions of co-operative abundance and altruism, there will— if conceptions of the good conflict—be a need for the fair allocation of benefits and burdens, for the assigning of duties and the protection of rights: but we should then need them in the face of the benevolence rather than the selfishness of others. Altruists, sincerely and conscientiously pursuing their respective conceptions of the good, could certainly cause injustice and violate rights. For every conception of the good favours certain social relationships and forms of life, and certain ways of defining individuals' interests—or, more precisely, certain ways of conceiving and ranking the various interests, deriving from their roles and functions, that individuals have. It also disfavours others. In a world in which no such conception is fully realized and universally accepted, even the non-egoistic practitioners of one threaten the adherents of others: hence the need for justice and rights.

But what if divergent conceptions of the good, and of interests, were to converge within a single moral and political consensus? Here a fourth condition comes into view: the lack of perfect information and understanding. For even under co-operative abundance, total altruism, and the unification of interests within a common conception of the good, people may get it wrong: they may fail to act as they should towards others, because they do not know how to or because they make mistakes, with resulting misallocations of burdens and benefits, and damage to individuals' interests. We may, therefore, say that if these four conditions obtain, a necessity exists for finding principles of justice for distributing social advantages and disadvantages, and principles specifying rights and duties, to protect us from one another's depredations and abuses, whether these be selfish or benevolent, intended or unintended.

Now, it is a peculiar and distinctive feature of marxism that it denies that these conditions of *Recht* are inherent in human life. *Recht*, Marx and Engels wrote, 'arises from the material relations of people and the resulting antagonism of people against one another' (Marx and Engels 1845–6: 318). Both could, and would, be overcome. Marxism specifically denies that scarcity, egoism, and social and moral antagonisms are invariant features inherent in the human condition, and it looks forward to a 'transparent' form of social unity, in which social life will be under the rational control of all. In short, it envisages the removal of the basic causes of significant conflicts of interest in society. As Marx and Engels wrote in *The Holy Family*,

If enlightened self-interest is the principle of all morality, man's private interest must be made to coincide with the interest of humanity. . . . If man is shaped by environment, his environment must be made human. (Marx and Engels 1845: 131, S. L.)

Marxism maintains that the conditions of *Recht* are historically determined, specific to class-societies, and imminently removable. Neither limits to desired goods, nor limited sympathies, nor antagonistic social relations, and corresponding moral ideologies, nor the opaqueness or reified character of social relations are essential to the human predicament. To assume that they are is itself an ideological illusion (propagated by *Recht*) — ideological in serving to perpetuate the existing class-bound social order. Marxism supposes that a transparent and unified society of abundance—a society in which the very distinctions between egoism and altruism, and between the public sphere of politics and the private sphere of civil society, and 'the division of the human being into a *public man* and a *private man*' (Marx 1843a: 155) have been overcome—is not merely capable of being brought about, but is on the historical agenda, and indeed that the working class is in principle motivated to bring it about, and is capable of doing so.

Thus *Recht* is not merely inherently ideological, stabilizing class societies and concealing class interests, and falsely purporting to adjudicate competing claims, limit freedoms, and distribute costs and benefits in a universally fair, objective, and mutually advantageous manner. It also presupposes an account of the conditions that call it forth that marxism denies. For marxists hold that, broadly, all significant conflicts are to be traced back to class divisions. So, for example, Marx and Engels could speak of communism as 'the *genuine* resolution of the conflict between man and nature and between man and man' (Marx 1844c: 296), and speculate about the abolition of crime under communism, and suggest that 'social peace' might succeed 'social war' (see Phillips 1980: ch. 4); and Trotsky, as we shall see, could proclaim that the future 'society without social contradictions will naturally be a society without lies and violence' (Trotsky 1938: 27). Certainly the marxist canon has virtually nothing to say about any bases of conflict, whether social or psychological, other than class. It is virtually innocent (and totally so at the level of theory) of any serious consideration of all the inter-personal and intra-personal sources of conflict and frustration that cannot, or can no longer, plausibly be traced, even remotely, to class divisions.

By furnishing principles for the regulation of conflicting claims and interests, *Recht* serves to promote class compromise and thereby delays the revolutionary change that will make possible a form of social life that has no need of *Recht*, because the conditions of *Recht* or the circumstances of justice will no longer obtain. In this respect, I think that Marx's view of morality as *Recht* is exactly parallel to his view of religion, concerning which he wrote: 'To abolish religion as the *illusory* happiness of the people is to demand their *real* happiness. The demand to give up illusion about the existing state of affairs is the *demand to give up a state of affairs which needs illusions*' (Marx 1844a: 176). Analogously, the demand to give up illusions about the 'rights of man' and 'justice' is the demand to give up the conditions of *Recht* and the circumstances of justice. Once emancipation from such conditions or circumstances arrives on the historical agenda, the morality of emancipation dictates the bringing into being of a world in which the morality of *Recht* is unnecessary. In that world, the conditions that made such a morality necessary will, as Engels put it, have been not only overcome but forgotten in practical life. This is the meaning of Lukács's accurate statement that the 'ultimate objective of communism is the construction of a society in which freedom of morality will take the place of the constraints of *Recht* in the regulation of all behaviour' (Lukács 1919b: 48).

One marxist thinker who made this pattern of thought particularly clear was the early soviet jurist Pashukanis. He saw private law, especially the law of contract, as the essential form of law, presupposing a world of self-interested, isolated, and competitive subjects, that is, possessors of commodities. Law, he thought, was 'not an appendage of human society in the abstract', but 'an historical category corresponding to a particular social environment based on the conflict of private interests' (Pashukanis 1924: 72). Thus, 'morality, law and the state are forms of bourgeois society', forms 'incapable of absorbing [a socialist] content and must wither away in an inverse ratio with the extent to which this content becomes reality'. He saw 'the social person of the future, who submerges his ego in the collective and finds the greatest satisfaction and the meaning of life in this act' as signifying 'the ultimate transformation of humanity in the light of the ideas of communism'. Morality was 'a form of social relation in which everything has not yet been reduced to man himself. If the living bond linking the individual to the class is really so strong that the limits of the ego are, as it were, effaced, and the advantage of the class actually becomes identical with personal advantage, then there will no longer be any point in speaking of the fulfilment of a moral duty, for there will be no such phenomenon as morality.' Morality, as *Recht*, will have withered away; but, like the founding fathers of marxism, Pashukanis saw a morality of emancipation—what he called 'morality' in 'the wider sense' —as 'the development of higher forms of humanity, as the transformation of man into a species being (to use Marx's expression)' (ibid.: 159–61).

The paradox with which we began is thus resolved, though not, I must admit, entirely. In particular, two difficulties or puzzles remain.

In the first place, nothing I have said accounts for Marx's aversion to 'prescribing recipes to the cookshops of the future' (Marx 1867 Afterword: 17, S. L.) or Engels's dismissal of imaginary constructions 'of an ideal society as perfect as possible'. I have explained why they rejected the morality of *Recht*, but why, having embraced the morality of emancipation, should they decline to specify its objectives, and more particularly its concrete practical and institutional implications for the future society?

Here we come upon a paradox within a paradox, a sub-paradox within the larger paradox we have been considering. For marxian and marxist thought is both anti-utopian and utopian. (By 'utopias' I here mean schemes of ideal societies that are doubly unrealistic, contrasting with existing evils by highlighting them, and unrealizable within existing parameters.) On the one hand, it has from the beginning always striven to differentiate itself from utopian socialism, claiming by contrast to be scientific and revolutionary; on the other, it has plainly centered on a vision of a future emancipated world that it holds to be latent in the womb of the present. In Marx and Engels, these two positions co-exist in constant tension with one another. In the subsequent marxist tradition they have separated out. The over-all drift of the mainline marxism of the Second and Third Internationals (and of Trotskyism and many social-democratic variants) has been scientific-anti-utopian; but utopian counter-currents have always existed, finding their most eloquent expression in the thought of Ernst Bloch. Bloch, as Hudson has remarked, drew attention to 'latent dimensions in Marx's thought and to under-emphasised texts' (Hudson 1982: 56), and to the integral link in marxian thought between the ability to understand the world and the will to change it. Bloch cited, in this connection, Engels's view that it was the vocation of socialists to become Templars of the Holy Grail and to risk their lives in a 'last holy war that will be followed by the milennium of freedom' (cited Hudson 1982: 61). For Bloch, marxism discovers 'concrete' rather than 'abstract' utopia in 'the not yet (*noch nicht*) actual objective real possibilities in the world'. Bloch seems to have believed that, through his economic analysis, Marx had uncovered such possibilities; thus his 'whole work serves the future, and indeed can only be understood and carried out in the horizon of the future, not indeed as one that is depicted in an abstract utopian way, but as one that takes effect in and out of the past as well as the present' (Bloch 1954–9: 727, 724–5).

How is this sub-paradox—marxism as utopia and anti-utopia—to be resolved? Superficially, we may say that marxian and marxist thought has always been ambivalent, that Marx was a utopian *malgré lui*, and that marxists have followed him in this, with more or less success at suppressing the utopian impulse. More deeply, we may say that what we have here is a coherent theoretical position, a kind of anti-utopian utopianism, distinctive of marxism. After arguing for each of these interpretations in turn, I shall suggest that marxism's anti-utopianism has weakened and subverted its utopianism, to the considerable detriment of marxism itself, both in theory and in practice.

Attacks by Marx and Engels and later marxists on utopians—notably Owen, Saint-Simon, and Fourier, but also the Babouvists and the Germans, notably Weitling and the hapless Herr Dühring—are well known. They painted 'fantastic pictures of future society, painted at a time when the proletariat is still in a very undeveloped state and has but a fantastic conception of its own position', and corresponding to 'the first instinctive yearnings of that class for a general reconstruction of society' (Marx and Engels 1848: 515). Marx's and Engels's main critical attention was focused on Owen, Saint-Simon, and Fourier, and it is as well to recall, with the Manuels, that their evaluations of these three, which range from contempt to generous praise, 'swayed with the subject under discussion and the political exigencies of the times' (Manuel and Manuel 1979: 702). What is worth accentuating here is the positive: the reasons for which they valued and were decisively influenced by all three.

Engels stated the most positive general appreciation in 1870:

German theoretical socialism will never forget that it rests on the shoulders of Saint-Simon, Fourier and Owen—three men who in spite of all their fantastic notions and all their utopianism, have their place among the most eminent thinkers of all times, and whose genius anticipated innumerable things the correctness of which is now being scientifically proved by us. (Engels 1875: 652.)

'We delight', he wrote elsewhere, 'in these stupendously grand thoughts and germs of thought that everywhere break out through their phantastic covering' (Engels 1880: 121); and in the *Communist Manifesto* Marx and Engels say of the famous trio that they 'attack every principle of existing society. Hence they are full of the most valuable materials for the enlightenment of the working class' (Marx and Engels 1848: 516). Marx saw their utopias as 'the anticipations and imaginative expression of a new world' (Marx 1866: 223).

In Owen they especially valued the practical determination to prove the virtues of co-operation, harnessing the abilities of all through the direct social organization of labour based on the factory system, and his commitment to the working class. Marx admired his 'really doughty nature' and revolutionary commitment (Manuel and Manuel 1979: 706), which Engels described thus:

Banished from official society, with a conspiracy of silence against him in the press, ruined by his unsuccessful communist experiments in America, in which he sacrificed all his fortune, he turned directly to the working class and continued working in their midst for thirty years. (Engels 1880: 127)

Indeed, Engels saw Owen's *New Moral World* as the most comprehensive project of the future communist community, with its 'groundplan, elevation, and bird's eye view' (ibid.: 126, S. L.). What united them with Saint-Simon and his followers has been well described by the Manuels: 'an endlessly dynamic prospect founded upon the boundless expansion of science and technology, exploitation of the inexhaustible natural resources of the globe, and the flowering of human capacities' (Manuel and Manuel 1979: 707). And from Fourier they took the notion of a community of richly diverse needs, as essential to many-sided individuality within a genuinely reciprocal community—though they balked at Fourier's interest in sensual and sexual experimentation. Marx the Victorian moralist was 'much more restricted in his outlook and recognised as legitimate only reasonable, refined and decent needs—which stopped far short of Fourier's equation of desires and needs' (ibid.: 710). In short, the Manuels are absolutely right to see these three utopian socialists as having 'left an indelible stamp on the banderole of the *Critique of the Gotha Programme*' (ibid.: 701), whose inscription was intended to describe communism's higher phase (Marx's and Engels's utopia): 'From each according to his abilities, to each according to his needs'.

So far I have sought to show that Marx and Engels were anti-utopian and utopian. They distanced themselves from the so-called 'Utopian Socialists' whose fantasies they saw as premature and pre-scientific, while synthesizing and incorporating their visions of the future into their own vision of emancipation. That vision, we may observe (following the most helpful classifications of J. C. Davis) embodies elements of Cockaygne, Arcadia, the perfect moral commonwealth or Kingdom of Ends, and the millennium (Davis 1981: 20–40).[2] How, then, could their anti-utopianism be reconciled with the embracing of this

---

[2] According to Davis, these are four types of ideal society which differ in their approaches to 'the collective problem: the reconciliation of limited satisfactions and unlimited human desires within a social context'. Thus,

utopian vision? To answer this, we must look more closely at the reasons for their anti-utopianism.

Marx and Engels were not opposed to utopianism in the sense of vesting high hopes in the future: few have held higher hopes than they did. But they did criticize the Utopian Socialists for drawing up utopian *blueprints*, just because in doing this they laid claim to a type of knowledge, social forecasting, that could not be had *now*: this was, as Lenin said, 'to indulge in idle guesswork about what cannot be known' (Lenin 1917a: 458). As Marx and Engels had written, 'the "whence" and the "whither" . . . are never *known* beforehand' (Marx and Engels 1845: 23). Engels made the same point in discussing the goal of abolishing the distinction between town and country:

To be Utopian does not mean to maintain that the emancipation of humanity from the chains which its historic past has forged will be complete only when the antithesis between town and country has been abolished; the utopia begins only when one ventures, 'from existing conditions', to prescribe the *form* in which this or any other antithesis of present-day society is to be resolved. (Engels 1872–3: 628.)

But secondly, they saw the very project of speculating about the ideal society as 'utopian' and thus as both anti-scientific and

---

'The milennarian shelves the problem by invoking a *deus ex machina*. Some force or agency from outside the system will miraculously alter its balance and functioning. In the perfect moral commonwealth tradition, the necessity for a prior change in the nature of men's wants is assumed. Men's desires become limited; limited, in fact, to the satisfactions, both material and sociological, that exist, as a whole and for particular groups and classes. In the Land of Cockaygne precisely the reverse happens. Material and sociological scarcities are wished away. Men's appetites remain unlimited but satisfactions multiply, on the material level especially, for their accommodation. Satisfactions are private rather than communal. Arcadia is found somewhere between these last two. Satisfactions are abundant but it is men's 'natural' rather than their conventional desires which are most apparent. Thus the Arcadian tampers with both sides of the collective problem.' (Davis 1981: 36–7.)

Readers of Davis will note that, in *his* sense of 'utopia' ('a holding operation, a set of strategies to maintain social order and perfection in the face of the deficiencies, not to say hostility, of nature and the wilfulness of men' which 'does not assume drastic changes in nature or man' [ibid.; 37]), Marx and Engels were not utopians. In *his* terms, I suggest that Marx and Engels were, predominantly, millennarian Arcadians, who envisaged the abundance, without the vulgarity and individualism, of Cockaygne, and a perfect moral commonwealth, in which, however, the morality of duty would wither away.

anti-revolutionary. Indeed, for them 'utopian' was an antonym of both 'scientific' and 'revolutionary': the early utopians had speculated about the ideal society just because they were not (and could not be) clearly aware of the future latent in the present, and so failed to identify with the class that would bring it into being. (Later utopias, such as those constructed by Dühring 'out of his sovereign brains', were simply 'silly, stale and reactionary from the roots up' [Marx 1877: 376].) As Marx wrote,

Just as the *economists* are the scientific representatives of the bourgeois class, so the *socialists* and the *Communists* are the theoreticians of the proletarian class. So long as the proletariat is not yet sufficiently developed to constitute itself as a class, and consequently so long as the very struggle of the proletariat with the bourgeoisie has not yet assumed a political character, and the productive forces are not yet sufficiently developed in the bosom of the bourgeoisie itself to enable us to catch a glimpse of the material conditions necessary for the emancipation of the proletariat and for the formation of a new society, these theoreticians are merely utopians who, to meet the wants of the oppressed classes, improvise systems and go in search of a regenerating science. But in the measure that history moves forward, and with it the struggle of the proletariat assumes clearer outlines, they no longer need to seek science in their minds; they have only to take note of what is happening before their eyes and to become its mouthpiece. So long as they look for science and merely make systems, so long as they are at the beginning of the struggle, they see in poverty nothing but poverty, without seeing in it the revolutionary, subversive side, which will overthrow the old society. From the moment they see this side, science, which is produced by the historical movement and associating itself consciously with it, has ceased to be doctrinaire and has become revolutionary. (Marx 1847a: 177–8.)

Here Marx is saying, in quasi-Hegelian fashion, that with the forward movement of history, a vantage-point becomes available from which the self-transformation of capitalism into socialism becomes increasingly visible. Adequate knowledge of this process, though not of the shape of the future society, becomes available to the scientific observer. And Marx wrote that the only appropriate response to such scientific knowledge is a revolutionary one: such an observer can only become a partisan theoretician of the proletarian class. Lukács, among later marxists, stated this aspect of Marx's thought with maximum force and clarity:

. . . since the ultimate objective has been categorised, not as Utopia, not as *reality which has to be achieved*, positing it above and beyond, the immediate advantage [of revolutionary classes and parties] does not mean abstracting from reality and attempting to impose certain ideals on reality, but rather it entails the knowledge and transformation into action of those forces already at work *within* social reality—those forces, that is, which are directed towards the realisation of the ultimate objective. Without this knowledge the tactics of every revolutionary class or party will vacillate aimlessly between a *Realpolitik* devoid of ideals and an ideology without real content. It was the lack of this knowledge which characterised the revolutionary struggle of the bourgeois class. . . . The Marxist theory of class struggle, which in this respect is wholly derived from Hegel's conceptual system, changes the transcendent objective into an immanent one; the class struggle of the proletariat is at once the objective itself and its realisation. (Lukács 1919a: 4–5.)

The knowledge that that theory expresses is the knowledge of a self-transforming present, not of an ideal future.

But, one must ask, how can one have the one kind of knowledge (of the self-transforming present) without the other (of the shape of future society)? How can scientific observers know that 'what is happening before their eyes', the result of 'forces already at work within social reality', is the 'realisation of the ultimate objective', the emancipatory transformation of capitalism into socialism, unless they also know, or at least have good reason to believe, that the 'new society', latent in the old, will take a form that *is* emancipatory, thus justifying their support for the proletariat's struggle? In other words, to assume that they do know the former is to assume that they know, or have good reason to believe, the latter.

Marx and Engels failed to see the force of this objection, for various reasons. If they had done so, they would have seen the inescapable need both to specify possible futures as closely as possible, indicating which are more or less probable, and to set out the grounds for supporting the struggle for one of them, by showing how it could realize values that would justify that support—a need that was all the greater since the future they envisaged was currently unfeasible and would benefit only future generations, thus requiring, as Plekhanov saw, self-sacrifice in the present. Instead of this, they fixed upon one such future, which they conceived as feasible, that is, realizable through revolutionary transformation, that was itself on the historical agenda, even imminent. They believed that its realization depended in part upon the prevalence within the revolutionary class of an anti-utopian mentality, which saw it as the coming future rather than as utopia; and that it could only be delayed by raising questions about its feasibility and desirability and only hastened by refusing to face them. Central to that anti-utopian mentality was a (utopian) redrawing of existing parameters, of the boundaries of 'realism', seeing the ideal future as *really* about to emerge from the present, on the assumption that, as a self-fulfilling prophecy, such a belief would help make such projections true.

They failed to see the objection and these corollaries for at least four reasons. First, they assumed a teleological philosophy of world history, alongside or rather behind their causal-cum-intentional explanations (see Elster: 1985), according to which the goal of world history is the resolution of all the antitheses of present-day society, and communism 'the complete return of man to himself as a *social* (i.e. human) being' and 'the riddle of history solved, and it knows itself to be this solution' (Marx 1844c: 296–7); and they interpreted world history and current events in the light of that goal.

It is true that, in criticism of Hegelian thought, and to some extent his own, Marx wrote that if 'later history is made the goal of earlier history', the 'real process is speculatively distorted'; and he added that 'what is designated with the words "destiny", "goal", "germ", or "idea" of earlier history is nothing more than an abstraction from later history, from the active influence which earlier history exercises on later history' (Marx and Engels 1845–6: 50). Indeed, Marx at one point explicitly said that 'communism as such is not the goal of human development' (Marx 1844c: 306).

But Marx and Engels never really shook off this mode of thinking. In a passage rather typical and highly revelatory of marxian and much later marxist thought, they wrote, in *The Holy Family*,

. . . the proletariat can and must emancipate itself. . . . It is not a question of what this or that proletarian, or even the whole proletariat, at the moment *regards* as its aim. It is a question of *what the proletariat is*, and what, in accordance with this *being*, it will historically be compelled to do. Its aim and historical action is visibly foreshadowed in its own life situation as well as in the whole organisation of bourgeois society today. . . . a large part of the English and French proletariat is already *conscious* of its historic task and is constantly working to develop that consciousness into complete clarity. (Marx and Engels 1845: 37.)

And consider, for example, Marx's notorious remarks about the British in India:

England, it is true, in causing a social revolution in Hindoustan, was actuated only by the vilest interests, and was stupid in her manner of enforcing them. But that is not the question. The question is, can mankind fulfil its destiny without a fundamental revolution in the social state of Asia? If not, whatever may have been the crimes of England, she was the unconscious tool of history in bringing about that revolution. Thus, whatever bitterness the spectacle of the crumbling of an ancient world may have for our personal feelings, we have the right, in point of history, to exclaim with Goethe:

> Sollte diese Qual uns qualen
> Da sie unsere Lust vermehrt,
> Hat nicht myriaden Seelen
> Timur's Herrschaft aufgezehrt?
>
> (Marx 1853: 351)

Thus Marx made the rash and strange assumption that 'mankind always sets itself only such tasks as it can solve' (Marx 1859: 363), supposing thereby that success in the task of building socialism was somehow historically guaranteed.

Second, Marx and Engels thought that the evils of capitalism were so blatantly obvious and the possible future so probable *and* obviously desirable that speculation about the latter was unnecessary. Just as emancipation had followed slavery and serfdom, so human emancipation would follow wage slavery. As he remarked in his Inaugural Address to the International Working Men's Association, 'like slave labour, like serf labour, hired labour is but a transitory and inferior form, destined to disappear before associated labour plying its toil with a willing hand, a ready mind and a joyous heart' (Marx 1864b: 383). To ask whether and why such emancipation was superior to what preceded it was, for Marx and Engels, absurd, a question that simply did not arise.

Third, they simply assumed that the working class was becoming the vast majority of capitalism's population, increasingly unified around a commitment to a socialist world-view, and to the realization of an emancipated future of communism. They assumed, therefore, that questions about the shape of future institutions and social arrangements would be faced and solved in common by the ascendant majority class, and that anticipating such solutions was premature and presumptuous. The assumption of imminent majority socialist unity could no longer of course be held by the time of Lenin, who, in its absence, on the threshold of revolutionary action, was forced into such anticipations in *The State and Revolution*, for which the marxist tradition had not prepared him and which it has been inhibited from pursuing since.

Fourth, they saw such speculation as counter-revolutionary: as open to endless possibilities of disagreement, wasting revolutionary energies, and inclining those who engage in it to fruitless appeals to society as a whole or to the goodwill of leaders and statesmen. Marx, it is said, wrote to the English positivist Beesly in 1869 that 'The man who draws up a programme for the future is a reactionary' (cited Manuel and Manuel 1979: 698); the revolutionary, by contrast, avoids such disagreements, concentrates his energies and focuses his political activity by 'taking note of what is happening before [his] eyes' and becoming 'its mouthpiece'.

These are poor reasons, and whatever weight they may understandably have had in the nineteenth century they altogether lack in the twentieth, in which 'what is happening before the eyes' of socialists has been the decline of the traditional working class and the persistent non-transformation of advanced capitalism into a higher form of emancipated society; the Stalinist route to socialism and the grim present-day experience of 'actually existing socialism' in Eastern Europe and parts of the Third World; and above all the general realization that mankind is beset by mutually contradictory tasks that it can barely comprehend, let alone know how to solve.

The sway of these reasons within the marxist tradition has consistently inhibited it from spelling out what the morality of emancipation implies for the future constitution and organization of society. As Irving Howe has well said, 'An intellectual scandal has been [socialism's] paucity of thought regarding the workings of socialist society: most marxists, in fact, have not thought it worth the trouble' (Howe 1981: 493). This inhibition has been damaging. Marxism has, in short, failed to exploit at least two of the real strengths of the utopianism it inherits and conceals.

In the first place, at the level of theory, utopian speculation has, since Plato's *Republic*, had a vital role to play in the clarification and elaboration of political and social ends through the imaginative exploration of the institutional and political forms that could embody them. Through counterfactual thought-experiments, basic questions about the ends of political life are pursued and answered in a distinctive and constructive way. As we shall see in Chapter Five, marxism has generally failed even to embark on this elucidatory enterprise.

Second, and far more damagingly, it has failed to exploit the practical strengths of utopian thinking, bringing liberating, non-routine perspectives to bear upon intractable problems and issues in the here and now. As we shall see, by trusting in what Lukács called the immanence of the ultimate objective, by believing that the (vaguely conceived) ends would call forth the appropriate means, it has almost totally failed to bring social and political imagination to bear upon the solution of real-life problems—such as the distribution of resources, social policy, economic, social, and industrial organization, political and constitutional structures, nationalism and regionalism. On such questions, marxism has had little distinctive and constructive to contribute: it has not been a *source* of creative solutions, a living tradition inspiring inventive policy-makers and planners. If the arguments I have advanced are right, the reasons lie deep within the structure of marxist thought itself.

The second remaining difficulty or puzzle concerning marxism's view of morality concerns its approach to justice. If the morality of *Recht* is ideological and imminently anachronistic, and if marxism adopts the morality of emancipation as its own, how are we to explain the fact that Marx and Engels sometimes appear to condemn the injustices and inequalities of capitalism;

and, more generally, what sense can we make of the very concept of exploitation, from which the notion of justice would seem to be inseparable? To this intriguing question the next chapter is in part addressed.

4   Justice and Rights

Justice

Did Marx think that capitalism, and more particularly the wage-relation between capitalist and worker, was unjust? A lively debate on this question has recently flourished, and by now all the logically possible positions on the issue have been ably and convincingly defended, viz.:

(1)   Marx thought the relation between capitalist and worker was just
(2)   he thought it was unjust
(3)   he thought it was both just and unjust—that is, just in one respect and unjust in another
(4)   he thought it was neither just nor unjust.

Position (1), which has come to be known as the Tucker-Wood thesis (see Tucker 1969 and Wood 1972, 1979, 1981), relies on a number of very telling passages. In *The Critique of the Gotha Programme*, in response to the Lasallean demand for 'a fair distribution of the proceeds of labour', Marx asks:

Do not the bourgeois assert that the present-day distribution is 'fair'? And is it not, in fact, the only 'fair' distribution on the basis of the present-day mode of production? Are economic relations regulated by legal conceptions or do not, on the contrary, legal relations arise from economic ones? (Marx 1875: 21.)

More specifically, he wrote that the capitalist's extraction of surplus value from the worker was 'by no means an injustice (*Unrecht*)' to the latter:

The circumstance, that on the one hand the daily sustenance of labour power costs only half a day's labour, while on the other hand the very same labour-power can work during a whole day, that consequently the value which its use during one day creates is double what [the capitalist] pays for that use, this circumstance is, without doubt, a piece of good luck for the buyer and by no means an injustice to the seller. (Marx 1867: 194, S. L.)

And in rebuttal of Adolph Wagner's suggestion that he, Marx, thought that the capitalist robs the worker, he wrote:

The obscurantist falsely attributes to me [the view] that 'the *surplus value* produced by the labourers *alone*, was left to the capitalist employers in an *improper way*'. Well, I say the direct opposite, namely, that commodity-production is necessarily, at a certain point, turned into 'capitalistic' commodity production, and that according to the *law of value* governing it, 'surplus value' is properly due to the capitalist and not to the labourer. . . .

. . . in my presentation, profit is *not* [as Wagner alleged] 'merely a *deduction* or "robbery" on the labourer'. On the contrary, I present the capitalist as the necessary functionary of capitalist production and show very extensively that he does not only 'deduct' or '*rob*', but forces the *production of surplus value*, therefore the deducting only helps to produce; furthermore, I show in detail that even if in the exchange of commodities *only equivalents* were exchanged, the capitalist—as soon as he pays the labourer the real value of his labour power—would secure with full rights, i.e. the rights corresponding to that mode of production, *surplus-value*. (Marx 1879–80: 186.)

In accordance with this last thought, Marx stated his more general position about the issue as follows:

The justice of the transactions between agents of production rests on the fact that these arise as natural consequences out of the production relationships. The juristic forms in which these economic transactions appear as wilful acts of the parties concerned, as expressions of their common will and as contracts that may be enforced by law against some individual party, cannot, being mere forms, determine their content. They merely express it. This content is just whenever it corresponds, is appropriate, to the mode of production. It is unjust whenever it contradicts that mode. Slavery on the basis of capitalist production is unjust; likewise fraud in the quality of commodities. (Marx 1861–79: iii. 333–4.)

And finally, he stated this position even more clearly in *Wages, Prices and Profit*:

To clamour for *equal or even equitable retribution* on the basis of the wages system is the same as to clamour for *freedom* on the basis of the slavery system. What you think just or equitable is irrelevant. The question is: what is necessary and unavoidable within a given system of production? (Marx 1865: 426. S. L.)

On the basis of passages such as these (all of which but the last he cites) Wood argues that transactions are just if they correspond or are appropriate to, or are functional to, the prevailing mode of production: judgements about justice are not made by reference to abstract or formal principles independent of the existing mode of production, indicating some ideal to which social reality could be adjusted; rather they are 'rational assessments of the justice of specific acts and institutions, based on their concrete functions within a specific mode of production' (Wood 1972: 16). Thus, since the exploitation of wage labour by capital is essential to the capitalist mode of production, there is nothing unjust about the transaction through which capital exploits labour; the worker is paid the full value of his labour power (unless, of course, he is defrauded), and the capitalist, in subsequently appropriating surplus value, is not required to pay the worker an equivalent for it, since under capitalism the worker has no right to the full value created by his labour. He did have such a right under the petty-bourgeois system of 'individual private property', but the very productive success of capitalism required its abolition. So, in short, according to Wood,

as Marx interprets it, the justice of capitalist transactions consists merely in their being essentially capitalist, in the correspondence of capitalist appropriation and distribution to those standards of justice which serve the system itself. (Wood 1979: 108.)

Capitalist exploitation 'alienates, dehumanises and degrades wage labourers', but 'it does not violate any of their rights, and there is nothing about it which is wrongful or unjust' (Wood 1981: 43).

Position (2), held by Husami, Cohen, and others relies on a variety of no less telling passages in which Marx plainly does speak of exploitation as 'robbery', 'usurpation', 'embezzlement', 'plunder', 'booty', 'theft', 'snatching', and 'swindling' (Husami 1978: 43–5). Thus, the 'yearly accruing surplus product [is] embezzled, because extracted without return of an equivalent, from the English labourer' (Marx 1867: 611, S. L.). So for example in the *Grundrisse* Marx wrote of

the theft of alien labour time [*i.e. of surplus value or surplus labour*] *on which the present wealth is based.* (Marx 1857–8: 705.)

And in *Capital* he wrote of the surplus product as

the tribute annually exacted from the working-class by the capitalist class. Though the latter with a portion of that tribute purchases the additional labour-power even at its full price, so that equivalent is exchanged for equivalent, yet the transaction is for all that only the old dodge of every conqueror who buys commodities from the conquered with the money he has robbed them of. (Marx 1867: 582.)[1]

In the light of passages such as these, Cohen maintains that

since, as Wood will agree, Marx did not think that by capitalist criteria the capitalist steals, and since he did think he steals, he must have meant that he steals in some appropriately non-relativist sense. And since to steal is, in general, wrongly to take what rightly belongs to another, to steal is to commit an injustice, and a system which is 'based on theft' is based on injustice. (Cohen 1983: 443.)

And Husami argues that those who defend Position (1) miss the satirical and ironic tone of the passages they cite in its defence, and that, by arguing that the only applicable standard of justice is that appropriate to the existing economic system, they make it 'impossible for the oppressed to criticise the injustice of their life situations' (Husami 1978: 52). According to Husami, Marx 'evaluates pre-communist systems from the standpoint of a communist society' (ibid.: 50): far from adopting capitalism's self-evaluation, he 'regarded capitalism as unjust precisely because, as an exploitative system, it does not proportion reward to labour contribution, and because it is not oriented to satisfy human needs' (ibid.: 78). This judgement, Husami insists, 'is made from the Marxian ethical standpoint which, Marx held, was a proletarian standpoint' (ibid.: 77).

---

[1] There is a further passage which tells heavily in favour of Position (2). In the *Grundrisse* Marx writes:

'The recognition [by labour] of the products as its own, and the judgment that its separation from the conditions of its realization is improper (*ungehörig*)—forcibly imposed—is an enormous awareness (*enormes Bewusstsein*), itself the product of the mode of production resting on capital, and as much the knell to its doom as, with the slave's awareness that he *cannot be the property of another*, with his consciousness of himself as a person, the existence of slavery becomes a merely artificial, vegetative existence and ceases to be able to prevail as the basis of production.' (Marx 1857–8: 463.)

Highly interestingly, in the 1961–3 *Critique*, Marx reproduces this passage but replaces '*ungehörig*' (improper) by '*ein Unrecht*' (an injustice) (Marx 1861–3: 6, 2287). I am grateful to Jon Elster for drawing this significant piece of evidence to my attention.

The defenders of Position (1) must somehow explain away the passages supporting Position (2), and vice versa. So it is no surprise to find Wood suggesting that Marx there uses 'robbery' in a special sense that does not imply injustice; and to find Husami stressing the 'ironic tone' of the cited passages concerning the justice of the capitalist wage relation. And indeed it *is* the case that the 'robbery' involved in capitalist exploitation has some special features, of which three are worth noting. First, as Marx himself observes, what the capitalist steals he himself 'helps create' by 'forcing the production of surplus value': if it were not for the capitalist, there would be nothing to rob the workers of. Second, as Marx elsewhere notes in criticism of Proudhon, their robbery is robbery according to bourgeois property rights, not necessarily according to other criteria. And third, as Wood points out, if the robbery relation is like that of conqueror to conquered, 'it is not so clear that robbery has to be unjust' (Wood 1981: 137–8), since it constitutes a regular production relation sanctioned by prevailing norms of justice. And it is also the case that some of the passages about the alleged justice of the capitalist's relation to the worker are highly satirical and ironic. On the other hand, Marx does plainly and frequently say that the relation is one of robbery, and he also plainly and occasionally says, sometimes that it is just, and sometimes that it is not unjust.

Position (3), which has been ably defended by Gary Young (Young 1975–6, 1978, and 1981), relies upon drawing a distinction, much favoured by Marx, between the sphere of exchange or circulation and that of direct production, and the correlative distinction between the worker as owner and seller of labour power, and the worker as 'a living component of capital', owned by the capitalist. Consider the following passages:

. . . the transformation of money into capital breaks down into two wholly distinct, autonomous spheres, two entirely separate processes. The first belongs to the realm of the *circulation of commodities* and is acted out in the *market-place*. It is the *sale and purchase of labour-power*. The second is the *consumption of the labour-power that has been acquired*, i.e. the process of production itself. . . . In order to *demonstrate* therefore, that the relationship between capitalist and worker is nothing but a relationship between commodity owners who exchange money and commodities with a free contract and to their mutual advantage, it suffices to isolate the first process and to cleave to its formal character. This simple device is no sorcery, but it contains the entire riddle of the vulgar economists. (Marx 1863–4: 1002.)

Secondly, the sphere of circulation or the exchange of commodities,

within whose boundaries the sale and purchase of labour-power goes on, is in fact a very Eden of the innate rights of man. There alone rule Freedom, Equality, Property and Bentham. . . . Equality, because each enters into relation with the other, as with a simple owner of commodities, and they exchange equivalent for equivalent. . . . On leaving this sphere of simple circulation or of exchange of commodities . . . we think we can perceive a change in the physiognomy of our dramatis personae. He, who before was the money-owner, now strides in front as capitalist; the possessor of labour-power follows as his labourer. The one with an air of importance, smirking, intent on business; the other, timid and holding back, like one who is bringing his own hide to market and has nothing to expect but—a hiding. (Marx 1867: 176.)

And finally,

It must be acknowledged that our labourer comes out of the process of production other than he entered. In the market he stood as the owner of the commodity 'labour power' face to face with other owners of commodities, dealer against dealer. . . . The bargain concluded, it is discovered that he was no 'free agent', that the time for which he is free to sell his labour-power is the time for which he is forced to sell it, that in fact the vampire will not loose its hold on him 'so long as there is a muscle, a nerve, a drop of blood to be exploited' [the quotation is from Engels]. (Ibid.: 301–2.)

Young criticizes Position (1) for failing to see the import of the distinction between worker as seller and worker as producer of surplus value, and insists that Marx thought both that the worker was treated justly (according to the laws of commodity exchange, specifying 'market rights'), and that the extraction of surplus value from him in the production process was robbery ('in the ordinary sense in which robbery is unjust' [Young 1981: 260]). In short, 'on Marx's view, the worker is treated justly as seller in the exchange of labour power for wages, but is then robbed in the production process, during which the capitalist extracts surplus value from the worker' (ibid.: 252). And Young further argues that, for Marx, only the latter is 'real', the former being merely ideological appearance, veiling and mystifying the transfer of surplus value, which is the essence of capitalist production.

Finally, Position (4), argued for by Richard Miller, relies on the observation that the passages on which Position (1) relies do not unambiguously support it. With regard to the wage relation, on the one hand their thrust is to deny that injustice is done, and on the other to insist that equivalents are exchanged. In these passages, the vocabulary of justice is used in a way that relativizes it to a mode of production and is, as we have seen, satirical, even ironic. ('Admire', he writes, 'this capitalistic justice!' [Marx 1867: 660]). Nowhere, as Miller writes, 'is there a non-relativized, unequivocal statement that capitalism is just. That is what one would expect if Marx does not regard justice as a fit category either for political recommendations or for scientific analysis' (Miller 1984: p. 80).

In other words, Position (4) focuses on Marx's view (see his attacks on Proudhon cited in Chapter Two) that justice is an archaic, scientifically irrelevant category, comparable to medieval theological notions; and on the view, which Marx also held, that its invocation is futile, and even dangerous, in social criticism and political action, in so far as it suggests objectively based and universally applicable standards for judging distributive arrangements and social institutions. Given this, 'the normal function of the term in criticism and justification should rationally be abandoned' (ibid.: 81). Appraising the relations of capitalism for their justice or injustice was scientifically anachronistic and politically fruitless.

What is one to make of this cacophony of interpretations? Some partial resolutions have been offered of some of the conflicting views we have sketched here. For example, Cohen (who adopts Position (2)) suggests that 'perhaps Marx did not always realise that he thought capitalism was unjust' (Cohen 1983: 444). More generally, Cohen has argued that:

Revolutionary marxist belief often misdescribes itself, out of lack of clear awareness of its own nature, and marxist disparagement of the idea of justice is a good example of that deficient self-understanding. (Cohen 1981: 12.)

Similarly, Jon Elster (who also adopts a version of Position (2)) suggests that 'both the theory of exploitation in *Capital* and the theory of distribution in *Critique of the Gotha Programme* embody principles of justice', but that 'like M. Jourdain, he did not know how to describe correctly what he was doing; unlike M. Jourdain, he actually went out of his way to deny that the correct description was appropriate' (Elster 1985). And Elster further suggests that 'the best way of making sense *both* of Marx's critique of capitalism and of the remarks on communism in the *Critique of the Gotha Programme* is by imputing to him a hierarchical theory of justice in which the contribution principle (to each according to his contribution) provides a second-best criterion when the needs principle (from each according to his ability, to each according to his needs) is not yet historically ripe for application' (ibid.).

Let us, then, look at what Marx explicitly says in *The Critique of the Gotha Programme* about the distributive arrangements of the future. Here if anywhere we should hope to find a clue to his positive thoughts about justice.

In communism's lower phase, he writes, 'still stamped with the birth marks of the old society from whose womb it emerges', each producer receives back from society means of consumption costing the same as the labour he has expended (minus various deductions for future investment, public services, and funds for those unable to work, etc.): '*equal right* here is still in principle—*bourgeois right*, although principle and practice are no longer at loggerheads' (Marx 1875: 23). But

this *equal right* is still constantly stigmatized by a bourgeois limitation. The right of the producers is *proportional* to the labour they supply; the equality consists in the fact that measurement is made with an *equal standard*, labour.

But one man is superior to another physically or mentally and so supplies more labour in the same time, or can labour for a longer time; and labour, to serve as a measure, must be defined by its duration or intensity, otherwise it ceases to be a standard of measurement. This *equal* right is an unequal right for unequal labour. It recognises no class differences, because everyone is only a worker like everyone else; but it tacitly recognises unequal individual endowment and thus productive capacity as natural privileges. *It is, therefore, a right of inequality, in its content, like every right. Right by its very nature can consist only in the application of an equal standard; but unequal individuals (and they would not be different individuals if they were not unequal) are measurable only by an equal standard in so far as they are brought under an equal point of view, are considered in one particular aspect only, for instance, as in the present case, are regarded only as workers and nothing more is seen in them, everything else being ignored.* Further, one worker is married, another not; one has more children than another, and so on and so forth. Thus, with an equal contribution of labour, and hence an equal share in the social consumption fund, one will in fact receive more than another, one will be richer than another, and so on. To avoid all these defects, right instead of being equal would have to be unequal. (Ibid.: 24, S. L., italics added.)

But what exactly are the defects? The first is simply that workers with higher productive capacities benefit by higher incomes. But this defect would simply be rectified by paying them all the same. Another is that some workers have dependents—members of their families—who do not work. But in an earlier paragraph of the *Critique* Marx has already allowed taxes to provide funds for those unable to work; and there seems to be no reason in principle why socialism could not provide non-working members of families with adequate incomes.

Aside from these particular and remediable defects, it is, however, the third defect, indicated by the passage in italics, which takes us to the heart of the matter. For here the objection is not that a particular principle of distribution is unfair, but rather that *any* system of rules specifying justifiable claims (*Recht*) treats people unequally, since, *by its very nature*, it applies a common standard to them, considering them in one particular aspect only. But, as Moore has pointed out, this amounts to a general argument that any social system is inequitable to the extent that it operates through general rules. According to this argument, 'no system of general rules, however complicated, can consider all the aspects in which individuals differ from one another. To apply such rules entails applying the same standard to different cases' (Moore 1980: 48–9). Marx would clearly not be satisfied by increasing the number of aspects in which people are considered, since his view appears to be that every respect in which individuals differ from one another could in principle be relevant; accordingly, *no* common standard could ever fit the bill. In short, he seems here to be taking all too seriously his doctrine of the 'universality of individual need, capacities ·. .', etc. and 'rich individuality that is as all-sided in its production as in its consumption'. He seems to have supposed that any rule of law or morals, which by its very nature singles out certain differences between people as grounds for differential treatment, is for that very reason 'abstract' and 'one-sided'. In the higher phase of communism, which the *Critique* goes on to describe,

after the enslaving subordination of the individual to the division of labour, and therewith also the antithesis between mental and physical labour, has vanished; after labour has become not only a means of life but life's prime want; after the productive forces have also increased with the all-round development of the individual, and all the springs of co-operative wealth flow more abundantly—only then can the narrow horizon of bourgeois right be crossed in its entirety. . . . (Ibid.: 24.)

I take this to mean, not merely that there will no longer be *bourgeois* right, but that there will be no more *Recht*, no more legal and moral rules: the horizon is a limit to thought and action set by bourgeois *Recht*; beyond it, there will be no bourgeoisie, and no *Recht*. The principle that such a society would inscribe on its banners—'From each according to his ability, to each according to his needs'—would not be such a rule, since (1) those abilities and needs would be infinite, that is, unlimitable in advance, and unspecifiable by any rule; (2) the former would be harnessed to 'the common interest of all individuals who have intercourse with one another' through what Marx called *gemeinschaftlich* relations (and Lenin 'a new and higher social bond'); and (3) the latter would all be satisfiable without conflicting claims because of those relations and because of material abundance.

I am now in a position to offer my own suggestion about how the dispute about Marx's views about capitalism's justice might be resolved. It starts from the observation that all four positions considered above are plausibly Marx's and are supported by textual evidence. Now, Cohen and Elster may be right in suggesting that Marx may, like M. Jourdain, have just failed to understand his own view of justice: I myself suggested in the previous chapter that he did not adequately reflect upon his own moral position. But I think we can go further in explicating the problem.

My suggestion is that Marx's view of capitalism's justice was both internally complex and hierarchically organized. In the first place, he did offer a functional account of the norms by which capitalist exploitation is judged just: the capitalist wage relation is judged just (on average, and apart from fraud, etc.) according to prevailing norms, viz. juridical norms of contract law, backed by conventions specifying the minimum socially necessary wage from the perspective of the vulgar economists, who 'translate the singular concepts of the capitalists, who are in the thrall of competition, into a seemingly more theoretical and generalised language, and attempt to substantiate the justice of those conceptions' (Marx 1861–79: iii. 226). These norms and the perspective prevail because they sanction and stabilize capitalist exploitation and thus the capitalist system. This is the truth in Position (1). Secondly, however, Marx also offered an 'internal' or 'immanent' critique of those norms and that perspective, as registering the mere appearance of an equivalent exchange of commodities. As Holmstrom has observed, it 'views the exchange between capitalist and worker too narrowly, abstracted from its background' (Holmstrom 1977: 366–7), failing to see that the worker is not free but 'forced' to sell his labour to (some) capitalist and thereafter under (that capitalist's) control. This is the truth in Position (3). But thirdly, Marx also offered an 'external' critique of capitalist exploitation and of the norms and perspective from which it appears just. That critique is in turn made from the perspective of communism's lower phase: capitalist exploitation is from this standpoint unjust because it violates the principle 'To each according to his labour contribution' (minus the appropriate deductions). This is the truth in Position (2). And finally, Marx offered a radical critique of capitalist exploitation, of the norms and perspective justifying it *and* of the critical perspective from which it appears unjust, from the perspective of communism's higher phase. From that standpoint, the very attribution of justice and injustice is a mark of class society, a sign that society is still in a prehistorical phase, an archaism eventually to be transcended. This is the truth in Position (4).

This solution to our interpretive puzzle may cause discomfort to someone who wants to know what, in the end, Marx actually believed *in propria persona*. Did he think capitalism unjust, or didn't he? But the answer, I believe, is that Marx maintained all these positions and that he brought all these perspectives to bear at once. So I disagree with Cohen's suggestion that Marx 'must have meant that the capitalist steals in some appropriately non-relativist sense', since, for Marx, there was no such sense: all such judgements are perspective-relative. Objectivity, in the sense of perspective-neutrality, was, for him, an illusion, indeed an ideological illusion.

I further disagree, as the whole argument of this book should make clear, with Elster's suggestion that the best way of making sense of both Marx's critique of capitalism and his vision of communism is to impute to him 'a hierarchical theory of justice', with the needs principle taking priority over the contribution principle. In my view, there is in Marx and marxism a hierarchy, but not a hierarchy of justice. I take a principle of justice to be one which is needed for 'assigning basic rights and duties and for determining . . . the proper distribution of the benefits and burdens of social co-operation', such that 'institutions are just when no arbitrary distinctions are made between persons in the assigning of basic rights and when the rules determine a proper balance between competing claims to the advantages of social life' (Rawls 1971: 5). What Marx offers is a multi-perspectival analysis in which capitalism's self-justifications are portrayed, undermined from within, and criticized from without, and then both justification and criticism are in turn criticized from a standpoint that is held to be beyond justice.

Exploitation

This way of interpreting Marx makes the best sense, I believe, of the marxist concept of exploitation, which, on this account, is internally complex and multi-perspectival, in exactly the way I have indicated. (I shall comment here only on capitalist exploitation.)

What, after all, are the defining features of capitalist exploitation, according to Marx? First, it is a market phenomenon. It is that form of surplus labour extraction in which, formally speaking, at the level of 'appearances', equivalent (labour power) is 'voluntarily' exchanged for equivalent (wages). Second, that extraction is achieved through the *power* of the exploiters (backed by that of the state): wage labour is forced labour, in which, under the 'dull compulsion of economic relations' (Marx 1867: 737)—as opposed to direct coercion—the labourer is, first, compelled to sell his labour power (though not to any given capitalist) and, second, then compelled to engage in the labour process under his master's supervision and control. The former compulsion is compatible with what Marx called 'formal freedom' (for instance, of whom to work for and what to purchase): it consists in the impersonal, anonymous constraints of the labour market, given the differential resources and organizational capacities of the agents of production, which render wage labour his only real option. The latter consists in the capitalist's legally-backed control—based on ownership of the means of production—over the labour process. Third, surplus value is extracted from the labourer *unfairly* or unjustly, when judged against the contribution principle: 'the greater part of the yearly accruing surplus product [is] embezzled, because abstracted without return of an equivalent' (ibid.: 611). Of course, as stated before, Marx recognized that the employer helps create what he embezzles, and indeed would probably not have done so but for the prospect of such embezzlement. This last fact, and the workers' acceptance of it as inevitable and of their rewards as just, result from the general capitalist mentality internalized by capitalists and workers alike. Marxian exploitation counts as such only when set against the external standard of justice as fairness provided by the contribution principle.

Finally, exploitation involves the inhuman character of the capitalist–worker relationship itself—and exchange relationships in general (this being Marx's earliest view of the matter): its calculative, instrumental nature, based upon the pursuit of conflicting interests by the parties to it, who view and treat one another and themselves in a manner incompatible with truly human relationships. This aspect of exploitation is well brought out in a striking passage from the *German Ideology*, where Marx writes of exploitation as a 'utility relation' in which

I derive benefit for myself by doing harm to someone else (*exploitation de l'homme par l'homme*): in this case moreover the use that I derive from some relation is entirely extraneous to this relation. . . . All this is actually the case with the bourgeois. For him only one relation is valid on its own account—the relation of exploitation; all other relations have validity for him only insofar as he can include them under this one relation, and even where he encounters relations which cannot be directly subordinated to the relation of exploitation, he subordinates them to it at least in his imagination. The material expression of this use is money which represents the value of all things, people and social relations. (Marx and Engels 1845–6: 409–10.)

And an early note, dating from 1844, makes even clearer how Marx saw this, the most general aspect of exploitation:

I have produced for myself and not for you, just as you have produced for yourself and not for me . . . [our production] is not *social* production. . . . Each of us sees in his product only the objectification of his *own* selfish need, and therefore in the product of the other the objectification of a *different* selfish need, independent of him and alien to him.
   As a man you have, of course, a human relation to my product: you have *need* of my product. Hence it exists for you as an object of your desire and your will. But your need, your desire, your will are powerless as regards my product. . . . The *social* relation in which I stand to you, my labour for your need, is therefore also a mere *semblance*, the basis of which is mutual plundering. (Marx 1844b: 225–6.)

In short, this aspect of exploitation is identified from the perspective of human emancipation, to which the next chapter is devoted.

Rights

Finally, we must ask: how do rights fit into the framework of Marx's and marxist thought? And, more particularly, can a marxist believe in human rights?

Plainly, Marx defended various particular rights in the course of his life, such as the right to a free press, the right to vote, workers' rights to decent factory conditions, and so on. The same goes for countless marxists ever since. Indeed, marxists across the world, especially since the Resistance to the Nazis, have been in the forefront of struggles against tyranny and oppression in many countries, often in the name of human rights, especially since the Helsinki Accords. In fact, I would argue that the establishment and protection of basic civil rights often depends on the existence of a strong and well-organized labour movement, and that marxist parties and groups have played an important role in achieving this.

Nevertheless, marxism as a body of thought has generally been inhospitable to rights. For instance, as Claude Lefort has remarked,

The expansion of marxism throughout the entire French left has for a long time gone along with a deprecation of law (*droit*) in general and the vehement condemnation, ironic or 'scientific', of the bourgeois notion of the rights of man. (Lefort 1981: 46.)

Rights are central to the theoretical tradition of liberalism (apart from its utilitarian strand), so that when they are violated in its name, this goes against the grain, so to speak, even if the violations are fully justified. Marxism, by contrast, displays no such tension. It is, as the next chapter will argue, from its origins committed to an ideal of freedom whose coming realization it labels human emancipation. It has never been similarly committed, as a matter of principle, to the promulgation and protection of *rights* that, when respected, serve to guarantee freedoms. For one thing, it has always tended to see rights as arising from and expressing the individualism and the contradictions of bourgeois society. For another, it is often ambivalent at best about the reality of bourgeois freedoms. And finally, it looks towards a future ideal society in which the freedoms it proclaims will require no guarantees.

A depressing illustration of what this way of thinking can lead to in practice is provided by the story of the early Soviet jurists Reisner, Stuchka, and Pashukanis. Accepting Lenin's view of the dictatorship of the proletariat as not supported by any laws, they used 'all their energy in order to prove the conservative function of the ideology of law', attacking law as such as conservative, ideological, tied to the 'commodity form', and as the 'opium of the people', and they justified the full subordination of law to politics in terms of 'revolutionary purposefulness' and 'political flexibility'. Pashukanis summed up this position when he wrote that 'revolutionary law' was 'ninety nine per cent a political task' (Tadic 1982: 424–5). As the Yugoslav *Praxis* philosopher Tadic has observed,

After the critical year of 1921, the dominant point of view in the Bolshevik party leadership theoretically advocated by Bukharin was that 'the party mechanism in the dictatorship of the proletariat can secure its leading position only under the condition of a monolithic unity. . . . This point of view led to the conclusion that the dictatorship of the proletariat was incompatible with political democracy . . . 'revolutionary purposefulness' soon turned into an arbitrary, despotic practice of 'state reason'. (Tadic 1982: 425.)

The roots of the marxist view of rights (and human rights) can be found in Marx's early essay 'On the Jewish Question'. There Marx made it plain that he saw rights—the rights of man of the American Constitution and of the French Declarations of 1789 and 1793—in one perspective only: 'nothing but the rights of a *member of civil society*, i.e. the rights of egoistic man, of man separated from other men and from the community' (Marx 1843a: 162). He saw them as securing the 'right of self-interest' of a monadic and 'restricted individual, withdrawn into himself and separated from the community' (ibid.: 163), and as basically reducing to the protection of private property; and as enshrining the illusory notion of a separate sphere of political emancipation (in which men *appear* free and equal) as a surrogate for (and precursor of) general human emancipation. In *The Holy Family*, he summarized his view even more extremely: 'It was shown', he wrote, 'that the *recognition of the rights of man* by the *modern state* has no other meaning than the *recognition of slavery* by the *state of antiquity* had' (Marx and Engels 1845: 133).

It must be said that Marx's was a narrow and impoverished view of the meaning of the rights of man, even in their late eighteenth-century forms: it treated them *only* as symptomatic of the individualism and contradictions of bourgeois life. Consider only some of the rights included in the 1789 Declaration (which became the Preface to the Constitution of 1791). Articles 10 and 11, for example, state:

No-one may be harassed for his opinions, even his religious opinions, provided that their expression does not disturb public order established by law.

and

The free communication of thoughts and of opinions is one of the most precious rights of man; every citizen can therefore speak, write and print freely, except that he may be prosecuted for the abuse of that liberty in cases determined by the law.

These rights cannot easily bear Marx's interpretation, even if opinion is seen as the private property of the individual. As Lefort has written, the second of these two articles make it clear that it is

the right of man, one of his most precious rights, to go beyond himself and relate to others, by speech, writing and thought. In other words, it makes clear that man could not be legitimately confined to the limits of his private wants, when he has the right to public speech and thought. Or in yet other words, for these last expressions risk reducing communication to the operations of its agents, individuals, defined one by one as examples of man as such, let us say that the article makes clear that there is a communication, a circulation of thoughts and opinions, of words and writing, which escapes in principle, except in cases specified by the law, the authority of power. The affirmation of the rights of man concerns the independence of thought and opinion in the face of power, and the cleavage between power and knowledge, not only and not essentially the divorce between bourgeois and citizen, between private property and politics. (Lefort 1981: 58–9.)

Or consider articles 7, 8, and 9:

No man can be accused, arrested or detained except in cases determined by law and according to the form it has prescribed. Those who solicit, expedite, execute or have executed arbitrary orders must be punished; but every citizen called or seized by virtue of the law must obey at once: he renders himself guilty by resistance.

and

The law can only establish punishments that are strictly and evidently necessary and no-one can be punished except by virtue of a law established and promulgated prior to the offence and applied legally.

and

Every man being presumed innocent until declared guilty, if it is judged indispensable to arrest him, every act of force that is not necessary to apprehend him must be severely punished by law.

These articles, all in the text Marx discusses, are passed over in silence. They do not lend themselves to his interpretation;

and he failed to consider their positive, world-historical significance, their applicability to non-egoistic, non-bourgeois forms of social life, and their consequent relevance to the struggle for socialism, because his mind was so exclusively fixed upon the critique of the egoism of bourgeois society and the mystifying ideology that pervaded it, from the perspective of a future he imagined as emancipated from both.

There is, of course, much to be said for that critique. It is full of insight and is still pertinent. In particular, it *is* probably true that the very concept of 'rights' is, in a sense, individualistic: property rights are centrally important in the contemporary liberal tradition and form a kind of paradigm for both its Lockean and Kantian variants. Rights are typically the basis for claims by individuals to be treated in certain ways: rights offer the *interests* of these individuals as sufficient grounds for holding another or others to be under an obligation to treat them in certain ways. It is probably true that an exclusively rights-based morality would be an impoverished one, unable to accommodate collective goods and the role of virtue in moral life: these are hard to capture in the form of individuated interests generating obligations (Raz 1984). But, as the passage quoted from Lefort eloquently shows, taking some rights seriously is positively to demand a certain form of social life in which social relationships flourish free of arbitrary political power. To think of them merely as expressing the egoism of civil society and the contradictions between civil society and the state is precisely to fail to take them seriously.

Moreover, the underlying thesis that we have attributed to Marx, that it is only the conflicting interests of class society that render rights necessary, is hard to believe in, in the face both of the historical record and of theoretical considerations. It is, after all, not only conflicting class interests that generate the need for the protections and guarantees that rights afford. As Buchanan has convincingly shown (Buchanan 1982), such protections and guarantees may be needed in at least the following circumstances: where minorities are disfavoured by democratic procedures, where paternalist policies interfere with individual liberties, where disagreements exist about what constitutes welfare or the common good, where coercion is required for the provision of public goods, and where guidelines and limits must be set to the provision for future generations. Why should we believe that any feasible form of social life, albeit abundant and free of class antagonisms, could be beyond circumstances such as these? Nor would the practice of altruism (which, as we have seen, Marx did not envisage) render such rights unnecessary. Even under altruism, there will be a need to protect people from others' mistakes about what altruism requires. Indeed, the more altruistic a society claims to be, the more important such protections and guarantees will be in case such claims are spurious, since only where they exist can the claims be tested.

My argument has been that marxism has inherited too narrow an account of the significance of rights and too narrow a view of circumstances that render them necessary. The former narrowness has made them seem unimportant, the latter potentially dispensable. But are there resources within the marxist tradition for overcoming this double narrowness? More specifically, can a marxist believe in human rights?

I have already said that countless marxists have in fact believed in and fought for human rights. So the question is not whether those whose beliefs and affiliations are marxist in fact believe in human rights. It is rather whether they can *consistently* do so. But the question thus formulated is still not adequate. For I am certain that many of those who are called, and call themselves, marxists and who believe in human rights hold a consistent set of beliefs that do not contradict their belief in and actions for human rights. The question, therefore, should be reformulated thus: can those whose beliefs and affiliations are marxist believe in human rights and remain consistent with central doctrines essential to the marxist canon?

To 'believe' in human rights, I shall assume, is, precisely, to take them seriously: to give priority to the interests they presuppose and the obligations they impose, and to be prepared to act accordingly when the occasion arises. Indeed, one test of such a belief is being prepared so to act.

Here I shall follow Feinberg in defining 'human rights' as 'generically moral rights of a fundamentally important kind held equally by all human beings, unconditionally and unalterably' (Feinberg 1973: 85). They are sometimes understood to be 'ideal rights', or rights that are not necessarily actually recognized, but which ought to be; that is, they ought to be positive rights and would be so in a better or ideal legal system. Sometimes, they are understood to be 'conscientious rights', that is, the claim is to recognize them as valid by reference to the principles of an enlightened conscience. Are they absolute?

To be absolute in the strongest sense, they would have to be absolutely exceptionless in all circumstances: given that the possessor of the right has the appropriate interests, the obligation the right imposes on appropriate others to protect or promote those interests would be categorical and absolute, admitting of no exceptions. But this is an impossibly strong requirement. For in the first place, the obligation, and thus the right, may be unfulfillable. This is the case with many of the rights specified in the United Nations Universal Declaration of Human Rights, especially the positive rights (rights to be treated in certain ways). These—for instance, the so-called 'social and economic rights', and in general rights to be given the means of living a decent life, or even a life at all—depend for their implementation on the availability of resources, and therefore, since ought implies can, they cannot be absolute in this sense. But second, and more fundamentally, the obligation, and thus the right, can always be legitimately overriden, in certain circumstances. Consider some promising candidates for absolute status. Take, first, the right not to be tortured. Suppose you have captured someone who knows where the bomb is that will blow up a crowded airport which there is no time to clear. The putatively absolute right not to be degraded or exploited falls to a similar objection. What of the more general rights to liberty, or to be treated with equal concern and respect? But what these mean in practice is unspecifiable in uncontestable terms; and in each case, it is not hard to think up non-eccentric situations in which they may be overriden for the sake of some greater good.

It is therefore perhaps better to say that human rights are strongly prima facie rights which, in general, are justified in defending people's vital interests and which, in general, override all other considerations bearing upon some policy or action, whether these concern goals and purposes or the protection of other, less central rights. They thus have a 'trumping' aspect (Dworkin: 1977): to believe in them is to be committed to defending them, even (or rather especially) when one's goals or strategies are not to be served, and indeed may be disserved, by doing so.

To put this another way (which shows the connection between justice and rights), talk of rights is a way of asserting the requirements of a relationship of justice, from the viewpoint of the persons benefiting from it. It involves adopting 'the viewpoint of the "other(s)" to whom something (including, *inter alia*, freedom of choice) is owed or due, and who would be wronged if denied that something' (Finnis 1980: 205). Talk of *human* rights is to do this, while emphasizing the fundamental and prima facie overriding status of this viewpoint with respect to certain matters, specifically those central to the flourishing of human beings. Proof that such talk is serious is being prepared to abandon goals and policies and strategies, except in rare and extreme cases, where to recognize such rights conflicts with their implementation.

On a narrower and more extreme view, rights might, following Robert Nozick, be seen as 'side constraints'—moral constraints upon goal-directed behaviour. This way of viewing rights (rather than building the minimization of the violation of rights into one's goals) is a strictly deontological position: the constraints must not be violated even if such violations would lead to better consequences (even if these consist in the minimizing of rights violations). Individuals, as Nozick puts it, 'have rights, and there are things no person or group may do to them (without violating their rights)'; their rights 'set the constraints within which a social choice is to be made, by excluding certain alternatives, fixing others, and so on' (Nozick 1974: ix and 166). This view of rights, according to Nozick, reflects the basic kantian position of treating persons as ends and not merely as means, of ruling out certain ways in which persons (or the Party or the State) may use others. On this view, their basis is that they

express the inviolability of other persons. But why may one not violate persons for the greater social good? Individually, we each sometimes choose to undergo some pain or sacrifice for a greater benefit or to avoid a greater harm: we go to the dentist to avoid worse suffering later: we do some unpleasant work for its results; some persons diet to improve their health or looks; some save money to support themselves when older. In each case, some cost is borne for the sake of the overall good.

Why not, *similarly*, hold that some persons have to bear some costs that benefit other persons more, for the sake of the overall social good? But there is no *social entity* with a good that undergoes some sacrifice for its own good. There are only individual people, with their own individual lives. Using one of these people for the benefit of others uses him and benefits the others. Nothing more. What happens is that something is done to him for the sake of the others. Talk of an overall social good covers this up. (Intentionally?) To use a person in this way does not sufficiently respect and take account of the fact that he is a separate person, that his is the only life that he has. He does not get some overbalancing good from his sacrifice. . . . (Ibid.: 32–3.)

But the trouble is that, for reasons stated above, it is not clear that any rights can be absolute. Perhaps, then, the constraints can be held to be unconstraining under certain circumstances? Nozick himself allows for this possibility when he writes in a concessionary footnote that 'the question of whether these side-constraints are absolute, or whether they may be violated to avoid catastrophic moral horror, and if the latter what the ensuing structure [of justification] might look like, is one I hope largely to avoid' (ibid.: 29–30, fn.). This 'threshold' approach requires some account of the conditions under which strongly prima facie rights, seen as constraints, may be overriden, and indeed of how rights differ in this regard; and this in turn would seem to require an analysis that is, in Sen's phrase, 'consequence-sensitive' (Sen 1981), comparing the consequences of obeying the constraint with those of violating it. Presumably such an analysis of *human* rights will set the threshold for such rights especially high.

Accordingly, Sen has proposed a consequence-sensitive approach to rights, building the fulfilment and non-realization of rights into the goals of action which is evaluated in terms of its consequences. This approach addresses directly the strength of rights claims against other considerations, especially where rights clash (e.g. where one right cannot be defended without another being violated). According to Sen, some rights are much more 'serious' than others, and these relate persons to basic capabilities to which they have a right (e.g. the general right not to be beaten up relates to the capability to 'move about without harm' [ibid.: 18–19]). So on this account too, rights set limits to other social policies and to the individual pursuit of goals (including the attainment of less important rights). *Human* rights, presumably, set maximally narrow limits.

Our question, then, reduces to this: has marxism any interest in taking such limits seriously, or is to do so not to take marxism seriously? My argument so far has tended to suggest the latter, first, because marxism sees them as expressive of the egoism of bourgeois society and, second, because it sees them as answering to a (pre-human) condition that must itself be transcended; *human* rights it tends to oppose as unwarrantably abstract and decontextualized. And yet at least one great marxist thinker has argued eloquently for a rights-based utopian perspective, namely Ernst Bloch, and the question inevitably arises: is not the future ideal of freedom to which marxism is committed unapproachable through the violation in the present and in the future of the limits that basic or human rights impose? (Of course, it might also be unapproachable through respecting them.) In the next chapter we shall look more closely at that ideal and its presuppositions; and in the final chapter at the question to which the foregoing discussion has inexorably led: what means are permissible in the course of its pursuit?

5 Freedom and Emancipation

One will have to reawaken in the breast of these people the sense of the self-worth of men—freedom. Only such a sense, which vanished from the world with the Greeks and evaporated into the blue with Christianity, can transform society again into a community of people for their highest ends—a democratic state.

(Marx 1843a: 25.)

Socialism, both in its ends and in its means, is a struggle to realise freedom.

(Korsch 1930: 126.)

It offends the human spirit to throw it into a destructive struggle unless it has a conception—if only of some essential features—of what might replace the world it is going to destroy.

(Kropotkin 1885: 308–9.)

How does marxism conceive of freedom, what does it promise emancipation from, and in what does that emancipation consist? This chapter seeks to answer these three questions.

The first is, one could say, a formal question, while the second and third are two sides of a substantive question. That is, the first asks: what manner of conceiving freedom is typical of marxism, as against other ideologies and belief systems? The second and third ask: what content does it give to its way of conceiving freedom, negatively and positively; how does it interpret the unfreedom of capitalism and the coming, real freedom of communism?

Marxism's conception of freedom

A common core on which all the richly diverse views of freedom or liberty can agree is this: that freedom is diminished when agents' purposes are prevented from being realized. This captures what we may call the *concept* of freedom—the non-constraining of the realization of agents' purposes—the interpretation of which is in dispute between the various *conceptions* of freedom (Rawls 1971: 5). Where these conceptions differ is indicated by the very formula just given. They diverge in three areas: first, in how to characterize the relevant purposes; second, in how to characterize the agent; and third, in how the freedom-diminishing constraints are to be understood. Moreover, their divergences on these three (related) matters are deep-lying and wide-ranging.

To begin with the constraints. The constraints that human beings face in their choice-making and actions can be classified in many ways (see Feinberg 1973). I shall here propose three. First, such constraints may be *external* (like handcuffs) or *internal* (like inhibitions). Of course, most real cases are mixed, such as legal rules, which, though externally given, are only constraining either because they are internally accepted or because of the internal fear of external sanctions. Notice that how this distinction is drawn is itself contingent upon how the agent is conceived: where the external-internal boundary lies depends on how the agent's self is conceptualized. Second, constraints may be *positive* (like prohibitions and taboos) or *negative* (like a lack of money or knowledge). Notice, again, that this is a shifting distinction in practice, since what counts as positive or negative depends on the description used (the presence of poverty is the absence of certain basic resources and life chances). Third, constraints may be *personal* or *impersonal* in origin: they may result directly from specified intentional acts by specific persons (as when a dictator imprisons me) or they may result from anonymous and impersonal processes and relationships (as when I cannot find a job). (Among personal constraints a further distinction may be drawn between those which are *intentionally* imposed and those which are not.) The distinction between personal and impersonal constraints may be a matter of degree (a constraint may be more or less directly related to intentional acts of persons); but it is also a shifting distinction in so far as views differ as to how, in turn, individual acts and impersonal relations and processes should be described and analysed. Some say the former should be seen only as instances of impersonal social processes; conversely, some say the latter must always be reducible to individual actions and interactions, and their consequences. (Neither of these extreme views is plausible.)

I turn next to that which the constraints constrain: the realization of agents' purposes. How are these to be individuated and identified? One answer is to tie them to *actual* desires or wants, or perhaps preferences, in choice situations: I am unfree if prevented from realizing these actual wants or preferences. This is a very restrictive view (on it a slave with few desires that are frustrated would be free). Another answer is to expand these to potential or counterfactual desires, wants, or preferences: I am unfree if I *would* be prevented from realizing these. On this view, what counts is the *range* of available purpose-realization, or want-satisfaction: how many doors are open? But both these are purely quantitative answers. Are all desires, wants, and preferences on a par, so far as freedom is concerned? Is my freedom to write and say what I please as much a freedom as my freedom to cross the road where I please? Indeed, is a restriction on the latter really a restriction of my *freedom*? So a further answer builds judgements of the significance of alternative purposes into deciding on their relevance to attributions of freedom: on this view, freedom and unfreedom only come into play when significant alternatives are subject to constraint. But how is significance decided? The natural answer is: by the agent. But, if this is so, on what basis? By the *urgency* of his wants, or the *intensity* of his desires, or by the exercise of *judgement*? Urgency and intensity seem very implausible: we are all familiar with urgently and intensely desiring what we know to be insignificant. But if it is a matter of the agent's *judgement*, is such judgement corrigible: can the agent be wrong about which of his purposes are significant (for him)? If not, is it really a matter of judgement (i.e. in which he *could* be right or wrong) rather than just *opting*, where there is nothing to be right or wrong about? And if it is corrigible, then what standards or criteria determine when it is based on 'confusion, illusion and distorted perspective' (Taylor 1979b: 192)? How is it to be ascertained which goals, aspirations, choices, actions, etc. are authentic, running with, rather than against, the grain of an agent's 'basic purposes' or self-realization? On this view, I am unfree if prevented from realizing my authentic or 'important purposes' (ibid.), those which are essential to me being the self-interpreting and self-realizing being that I am.

Which leads me to the third area of dispute: the nature of the agent. On the simplest and narrowest view, an agent to whom freedom or unfreedom is attributed is a mere locus of (actual or potential) wants and preferences: 'he' is more or less free as these are free from the pertinent constraints on their realization. On a more complex and wider view, the question of *autonomy* is introduced: are his wants or preferences genuinely 'his', autonomous rather than heteronomous, self-directed rather than imposed or induced? (Am I unfree if prevented from doing what a hypnotist commands?) This is, of course, a matter of degree; and, once more, this is a shifting distinction in practice, depending on how the relation between my reasons and my actions is understood (what makes that relation count as autonomous? Causing my action in 'the right way'? But what does that mean?) Enriching this view involves introducing the notion of '*authenticity*'. Agents, seen thus as self-interpreting and self-realizing subjects, typically engage in 'strong evaluation', preferences about preferences, making judgements about what sort of a person to be. On this view, 'we experience our desires and purposes as qualitatively discriminated, as higher or lower, noble or base, integrated or fragmented, significant or trivial, good or bad' (ibid.: 184). Freedom consists in not being impeded in one's authentic, or basic, purposes by obstacles which include inauthentic or 'repudiated' preferences.

A further crucial step, taken by Kant, involves distinguishing between a 'real' or rational and an 'unreal' self, or even, more contentiously still, a 'higher' and a 'lower' self, the first in each case being the genuine locus of freedom. With this step, a kind of *Gestalt* switch occurs, in which one's whole perspective on freedom shifts decisively. Freedom for the rational agent becomes freedom from heteronomy, from subjection to empirically caused desires: 'such independence', Kant wrote, 'is called freedom, in the strictest, i.e., transcendental sense' (*Critique of Practical Reason*, book 1, section 5). This is the freedom of the pure autonomous rational will, acting rightly according to the purely formal moral law it gives to itself. Constraints upon such freedom are not obstacles to want-satisfaction, or limits upon choice, but all (including natural desires and inclinations) that hinders a moral life based in pure reason. Freedom is rational self-determination.

A final step (and none of these steps is compelled by any earlier step) identifies the higher or real self with some collectivity or organization or community—a state, a class, a party, a nation— held to be constitutive of the agent's essential identity. On this last view, the freedom consists in the non-constraining of *that* agent's essential purposes: freedom, as Fichte wrote of the nation, 'from intrusion and corruption by anything alien' (cited Krieger 1957: 190). Freedom is collective self-determination.

Hegel took both steps, seeking to give Kantian freedom a social and political embodiment: seeing that freedom which 'finds its exercise in the sphere of particular and limited desire' as 'mere caprice', and contrasting it with the union of the *subjective* with the *rational* will':

it is the moral whole, the *State*, which is that form of reality in which the individual has and enjoys his freedom; but on the condition of his recognising, believing in and willing that which is common to the whole. (Hegel 1822–31: 38.)

Others, such as Fichte and other German nationalists, also took both these last steps, with implications unsurpassably described by Sir Isaiah Berlin:

Fichte knows what the uneducated German of his time wishes to be or do better than he can possibly know for himself. The sage knows you better than you know yourself, for you are the victim of your passions, a slave living a heteronomous life, unable to understand your true goals. You want to be a human being. It is the aim of the state to satisfy your wish. 'Compulsion is justified by education for future insight'. The reason within me, if it is to triumph, must eliminate and suppress my 'lower' instincts, my passions and desires, which render me a slave; similarly (the fatal transition from individual to social concepts is almost imperceptible) the higher elements in society—the better educated, the more rational, those who 'possess the highest insight of their time and people'—may exercise compulsion to rationalise the irrational section of society. (Berlin 1958: 149–50.)

Marx did not take either of these steps: he did not accept the kantian notion of the rational moral will and purely formal moral law; nor did he see individuals as freely self-determining only qua members of identity-constituting collectivities. Yet he was undoubtedly influenced by the fact that, within the German culture of his time, so many had taken these steps: self-determination was a decisively important idea for him, even if he never thought through to the end his notion of the 'self', the subject of constraints and of potential self-determination, or explained what emancipated self-determining agents would look like. The reason he did not do so is that he scarcely even addressed the question of identity, or self-definition, nor did he see it as a question to which human beings urgently require an answer. The need for definition and recognition within a particular community formed no part of Marx's philosophical anthropology: hence the peculiar incapacity of marxian and marxist thought to account for ethnicity, nationalism, regionalism, religion, and other identity-constituting cultural phenomena (other than as illusory, ideological products of class societies).

On the other hand, he undoubtedly rejected the narrow construal of liberty typical of that strand of liberalism which stretches from Hobbes and Hume through Bentham and Mill to the consensus dominant in Anglo-American thought and practice today. According to this, there is a tendency to see the constraints that bring unfreedom as external, positive, and personal; what they constrain as (actual or potential) wants or preferences, whatever they may be, irrespective of significance; and the agents or subjects of freedom as merely the loci of such wants and preferences. (This explains why liberals are often drawn to an 'opportunity' rather than an 'exercise' conception of freedom, according to which my freedom is solely a matter of how many doors are open to me. For opportunities are external to agents and can be identified without reference to them or their purposes.) This would be a crude account of liberalism in general, because plainly elements of a wider and richer view of liberty have been important to various liberal thinkers, notably Humboldt and J. S. Mill himself, who laid much stress on the strong evaluation of alternative life styles (disputing Bentham's view that pushpin is as good as poetry), and undoubtedly saw autonomy, authenticity and self-realization as essential to a free agent (comparing unfavourably a Socrates dissatisfied with a pig satisfied).

Nevertheless, it is not inaccurate to see much of liberalism—and certainly that central strand leading from Hobbes through Bentham to utilitarianism—as exemplifying the picture I have sketched. And indeed, Marx criticized utilitarianism (he had in mind Helvétius, Holbach, Bentham, and James Mill) on various grounds, but among them for its indiscriminate aggregative approach, for being exclusively preoccupied with the aggregation of an alleged single quality, utility (or, as we might now say, want-satisfaction), rather than attending to the *meaning* of actions and social relations, or the significance of the wants to be satisfied, which preoccupation he in turn related to the nature of capitalist social relations:

The apparent absurdity of merging all the manifold relationships of people in the *one* relation of usefulness, this apparently metaphysical abstraction arises from the fact that in modern bourgeois society all relations are subordinate in practice to the one abstract monetary-commercial relation. . . . the actual relations [of] speech, love, definite manifestations of definite qualities of individuals . . . are supposed not to have the meaning *peculiar* to them but to be the expression and manifestation of some third relation attributed to them, the *relation of utility*. . . . (Marx and Engels 1845–6: 409.)

More generally, much of the liberal tradition has purported to favour social and political arrangements that are neutral with regard to individual plans of life or conceptions of the good, focusing on not restricting (external) opportunities and presupposing a choosing self that is left free to choose its conception of the good. Hence, there has undoubtedly been what we may call a pressure within the liberal tradition to a narrow construal of the various components of liberty I have isolated. The most narrow construal of all is to be seen in F. A. Hayek's definition of liberty as 'the state in which a man is not subject to coercion by the arbitrary will of another or others' (Hayek 1960: 11).

In striking contrast, marxism falls, along with the thought of Spinoza, Rousseau, Kant, and Hegel, and others, into what I have called the category of wider, more complex, or richer views of freedom (and into the pitfalls and paradoxes they bring with them). It allows that the constraints on 'real freedom' may be internal, negative, and impersonal, it certainly discriminates between different interests and preferences with regard to their relevance to freedom and unfreedom, and it is fully committed to a conception of the agent as a (potentially) self-directing being who achieves self-realization in mutual identification and community with others. Thus, it is only too obvious, Marx allows in ideological, cognitive, and motivational inhibitions as constraints on freedom and, as we have seen, speaks of the dull compulsion of impersonal economic relations. He certainly distinguishes between significant and insignificant, authentic and inauthentic needs and wants (writing, for example, of the bourgeois's 'boundless greed after riches' [Marx 1867: 153]). And Marx's conception of the agent is indicated by passages such as this (in criticism of Adam Smith): Smith failed to realize that

[the] overcoming of obstacles is in itself a liberating activity—and that, further, the external aims become stripped of the semblance of merely external urgencies, and become posited as aims which the individual himself posits—hence as self-realisation, objectification of the subject, hence real freedom. . . . (Marx 1857–8: 611.)

By implication, therefore, marxism condemns the views we have referred to above both as narrow and as ideological in their narrowness, stabilizing class relations, and serving class interests by making it appear that there is far less unfreedom about than there actually is.

And yet it is one thing to reject liberal views of freedom as too narrow; it is another to deny bourgeois freedoms the status of genuine freedoms. And sometimes Marx and often subsequent marxists have done this. We have seen that in *On the Jewish Question*, Marx linked the right to liberty with egoism, self-interest, and private property. Sometimes he was concerned to specify the ideological understanding of 'freedom' prevalent within bourgeois relations of production, and to locate that understanding historically, by contrast both with previous production relations and with the projected post-bourgeois future:

By freedom is meant, under the present bourgeois conditions of production, free trade, free selling and buying. . . . This talk of free selling and buying, and all the other 'brave words' of the bourgeoisie about freedom in general have a meaning, if any, only in contrast with restricted selling and buying, with the fettered traders of the Middle

Ages but have no meaning when applied to the communistic abolition of buying and selling, of the bourgeois conditions of production, and of the bourgeoisie itself. (Marx and Engels 1848: 499–500.)

But in the *Grundrisse* he went further, describing freedom of competition as

nothing more than free development on a limited basis—the basis of the rule of capital. This kind of individual freedom is therefore at the same time the most complete suspension of all individual freedom, and the most complete subjugation of individuality under social conditions which assume the form of objective powers, even of overpowering objects—of things independent of the relations among individuals themselves. (Marx 1857–8: 652.)

And elsewhere, pursuing the same thought, if less extremely, he remarked that:

In imagination, individuals seem freer under the dominance of the bourgeoisie than before, because the conditions of life seem accidental; in reality, of course, they are less free, because they are to a greater extent governed by material forces. (Marx and Engels 1845–6: 78–9.)

More specifically, Marx spoke of the freedom (along with the equality) of the worker, in his exchange relationship with the capitalist, as being merely 'formal', masking his dependence on capital as a whole and the capitalist class in general. The 'legal relation' or '*ficto juris*' (Marx 1867: 547) of contract, in which the parties confront each other merely as individuals, is 'a mere *semblance*, and a *deceptive semblance*': the worker

sells the particular expenditure of forces to a particular capitalist, whom he confronts as an independent *individual*. It is clear that this is not his relation to the existence of capital as capital, i.e. to the capitalist class. Nevertheless, in this way everything touching on the individual, real person leaves him with a wide field of choice, of arbitrary will, and hence of formal freedom. (Marx 1857–8: 464.)

This mode of thinking lies at the root of a historical tendency within marxism to disparage as merely 'formal' the typical bourgeois freedoms (including civic and political freedoms), and from this disparagement it is a short step to deny them the status of genuine freedoms at all (a step which, as we have seen, Marx himself occasionally took). But, as the last quotation suggests, Marx's own considered view was more complex. There *was*, after all, a 'wide field of choice' left to the individual, both as worker, choosing how to deploy his abilities, whom to work for, etc., and as consumer, converting the money he earns 'into whatever use values he desires':

he acts as a free agent; he must pay his own way; he is responsible to himself for the way he spends his wages. *He learns to control himself, in contrast to the slave*, who needs a master. (Marx 1863–4: 1033.)

And indeed, is not the *reality* of the formal freedoms (as well as of the expression of opinion, and of political activity) essential to the workers' becoming capable of the revolutionary transformation that will bring into existence the 'real freedom' or emancipation of which Marx and so many subsequent marxists so eloquently speak (see Elster: 1985)? Let us now turn to examine what that reality comprises.

Alienation

First, we must ask: what is emancipation emancipation from? The most fundamental general answer to this question is: alienation. Marx elaborated his concept of alienation in his early writings as a wide-ranging idea whose purpose was to explain and connect together the many evils and irrationalities of modern society. I agree with Wood that in his later writings it is

no longer explanatory; rather it is descriptive or diagnostic. Marx used the notion of alienation to identify or characterize a certain sort of human ill or dysfunction which is especially prevalent in modern society. This ill is one to which all the various phenomena exemplifying the images or metaphors of 'unnatural separation' or 'domination by one's own creation' contribute in one way or another. (Wood 1981: 7.)

Of course, Marx attributed a number of ills to capitalism: among them class domination and exploitation, waste of resources and energies, irrationality, inefficiency, poverty, degradation, and misery. 'Alienation', however, captures those factors— particularly acute under capitalism—that constitute unfreedom, and whose abolition would constitute human emancipation. The other ills I have mentioned are, of course, not unrelated to unfreedom, but alienation captures the central obstacles to 'real freedom'. In short, 'alienation' is the name for what is distinctive of the marxian view of capitalist unfreedom, and for what, according to that view, makes capitalism distinctively unfree. In seeking to clarify what it refers to, I shall discuss, first, the main symptoms of this (terminal) illness; second, the nature of the illness; and third, its causes—according, that is, to Marx.

The most basic among the forms taken by alienation is the alienation of labour. This consists in

the fact that labour is *external* to the worker, i.e. it does not belong to his intrinsic nature; that in his work, therefore, he does not affirm himself but denies himself, does not feel content but unhappy, does not develop freely his physical and mental energy but mortifies his body and ruins his mind. The worker therefore only feels himself outside his work and in his work feels outside himself. He feels at home when he is not working and when he is working he does not feel at home. (Marx 1844c: 274)

The worker's alienation consists partly in the 'one-sided, crippled development' of 'increasingly unskilled labour' (Marx and Engels 1845–6: 262.)—the 'impoverishment of the labourer in his individual productive powers' (Marx 1867: 361, S.L.), so that the division of labour 'attacks the individual at the very roots of his life' (ibid.: 363), partly in alienation from his fellow-workers, no less impoverished and stunted by the division of labour, partly in alienation from the earth and the instruments of production which are 'the property of another' (Marx 1857–8; 497–8). The worker is also 'related to the product of his labour as to an alien object': thus

the more the worker spends himself, the more powerful becomes the alien world of objects which he creates over and against himself, the poorer he himself—his inner world—becomes, the less belongs to him as his own. . . . The *alienation* of the worker in his products means not only that his labour becomes an object, an *external* existence, but that it exists *outside* him, independently, as something alien to him, and that it becomes a power on its own confronting him. It means that the life which he has conferred on the objects confronts him as something hostile and alien. (Marx 1844c: 272.)

More generally, life under capitalism is alienated, for members of all classes, by virtue of the one-sided and compulsive character of the needs it encourages, the predominantly instrumental, mutually indifferent, non-reciprocal, and conflictual character of its social relations, and the fact that these relations themselves acquire a 'fetish-like' form (Marx 1861–79; iii. 391), a 'fetishism which metamorphoses the social, economic character impressed on things in the process of social production into a natural character stemming from the material nature of those things' (ibid.: ii. 225). Thus under captialism, 'inhuman, sophisticated, unnatural and *imaginary* appetites' are encouraged (Marx 1844c: 307), above all an obsessive need for money, so that 'all passion and all activity must be submerged in *avarice*' (ibid.: 309). Man becomes a 'debased, enslaved, foresaken, despicable being' (Marx 1844a: 182), whose 'life-activity, his *essential being*, [becomes] a mere means to his existence' (Marx 1844c: 276), and labouring activity is 'not the satisfaction of a need', but merely a 'means to satisfy needs external to it' (ibid.: 274). The typical social relations and institutions of capitalism—the exchange of commodities, capital, money—appear to individuals 'as an alien force existing outside them, of the origin and goal of which they are ignorant, which they thus are no longer able to control, which on the contrary passes through a peculiar series of phases and stages independent of the will and action of man, nay even being the prime governor of these' (Marx and Engels 1845–6: 48).

These are the main symptoms of alienation. One question, unasked by Marx, is whether the patient must be aware of his symptoms, whether they must be subjectively experienced as alienating for alienation to exist, or whether alienation may merely be objective, an observer's category, applied to both the situation and the self-understanding of the alienated. In the first case, 'alienation' captures phenomenologically the 'life-world' of the alienated, his world from within; in the second, it captures that world, and his experience of it, from the external perspective of a counterfactual world which his very alienation prevents from becoming factual. Marx never really addressed this problem—chiefly, I suppose, because he just assumed that objective alienation either was, or would inevitably become, subjective. But the question is important, for the answer to it bears directly upon the capacity of the alienated to perceive and to seize (indeed, even to want to perceive and to seize) whatever possibilities of emancipation may exist. And of course, it became especially important from the time of Lukács who, facing the problem Marx never addressed, was the first marxist thinker to take persisting objective alienation seriously.

What, then, were these various symptoms symptoms of? In what does alienation consist? The answer is, I think, twofold: alienation can be seen as having two (interrelated) aspects.

On the one hand, it can be seen as the suppression, or stunting, of possibilities of human development. Under alienation, individuals fail to 'affirm', 'confirm', and 'actualise' themselves (Wood 1981: 21). Marx held that it is 'the vocation, designation, task of every person to achieve all-round development of all his abilities' (Marx and Engels 1845–6: 292), and (in Marx's early formulation) when alienated, people's lives are not such that the human essence 'feels itself satisfied' (ibid.: 58), or (in a later formulation) their labour's external purposes appear as 'mere natural necessity' rather than the individual's own— as 'self-realisation, objectification of the subject, hence real freedom, whose action is, precisely, labour' (Marx 1857–8: 611).

Plamenatz has called this 'spiritual alienation' (Plamenatz: 1975)—and it is, as I have said, crucially ambiguous as between a subjective condition and an objective lack. It involves a number of elements most of them already alluded to: life is 'robbed of all real life content' (Marx and Engels 1845–6: 87) and rendered 'valueless' and 'unworthy' (Marx 1844c: 273); the division of labour results in 'the one-sided development of one quality at the expense of all the rest' (Marx and Engels 1845–6: 262) and the absence of true communal reciprocity, in which 'production as human beings' would imply that 'in the individual expression of my life I would have directly created your expression of your life, and therefore in my individual activity I would have directly *confirmed* and *realised* my true nature, my *human* nature, my *communal nature*' (Marx 1844c: 227–8). Moreover, the products of labour do not 'objectify' or express their lives as humanly meaningful: they are not 'so many mirrors in which we saw reflected our essential nature' (ibid.: 228) so that man 'sees himself in a world he has created', objectifying 'man's species life' (ibid.: 277). In short, capitalist methods 'mutilate the labourer into the fragment of a man', alienating from him 'the mental and spiritual potentialities of the labour process' (Marx 1867: 645); and more generally, the individual lacks 'social scope for the vital manifestation of his being' (Marx and Engels 1845: 131), and all, capitalists and workers, lack the kind of mutual association between individuals (assuming the advanced stage of modern productive forces, of course) which puts the conditions of 'the free development and movement of individuals under their control'—conditions which were previously left to chance and which had acquired an 'independent existence over against the separate individuals' (Marx and Engels 1845–6: 80).

The second aspect of alienation consists in what stunts or suppresses that free development: the distinctively marxian constraint on 'real freedom'. This is impersonal and anonymous, the 'consolidation of what we ourselves produce into a material power above us, growing out of our control', whereby 'man's own deed becomes an alien power opposed to him, which enslaves him instead of being controlled by him' (ibid.: 47–8). Thus under capitalism, 'for the proletarians . . . the conditions of their life, labour and with it all the conditions of existence of modern society, have become something extraneous, something over which they, as separate individuals, have no control, and over which no *social* organisation can give them control' (ibid.: 79).

Plamenatz and others call this 'social alienation', and it takes a number of forms: the imperatives of capitalist competition; the rigours of the labour market; the role requirements of the division of labour; the 'illusory community in which individuals have up till now combined' which 'always took on an independent existence in relation to them', since it concealed class domination and was thus 'for the oppressed class not only a completely illusory community, but a new fetter as well' (ibid.: 78); and the *fetishism* of commodities (in which 'the social character of men's labour appears to them as an objective character stamped upon the product of that labour' [Marx 1867: 72], relating not human beings but their products), of money and of capital. In short, reified social relations, constraining the real freedom constituted by many-sided self-development in community with others. These relations

can only be abolished by individuals again subjecting these material powers to themselves and abolishing the division of labour. This is not possible without the community. Only within the community has each individual the means of cultivating his gifts in all directions; hence personal freedom becomes possible only within the community. (Marx and Engels 1845–6: 77–8.)

What is the etiology of this illness: where do its principal causes lie? At one level, the answer is given within historical materialism. The relations that inhibit and preclude real freedom are precisely those social relations men enter into in 'the social production of their life', those 'definite relations that are indispensable and independent of their will, relations of production which correspond to a definite stage of development of their material productive forces' (Marx 1859: 362–3), along with the legal, political, and indeed moral superstructure that rises upon and stabilizes those very relations. Marx and later marxists have made a point of arguing that the immense growth of the productive powers of capitalism has occurred 'at the cost of the individual labourer' (Marx 1867: 645), such growth being both the means and the precondition for the eventual flowering of all individuals' species powers. Up to now, development of the latter has been both partial and class-related:

All emancipation carried through hitherto has been based . . . on restricted productive forces. The production which these productive forces could provide was insufficient for the whole of society and made development possible only if some persons satisfied their needs at the expense of others, and therefore some—the minority—obtained the monopoly of development. (Marx and Engels 1845–6: 431–2.)

The idea here seems to be that, before abundance, zero-sum class relations are inevitable: the (limited) historically available possibilities for self-development will be seized by some, at the expense of others.

At a deeper level, Marx's explanation for alienation goes back to Hegel and to Rousseau. Alienation enters human history at the point where human beings can no longer successfully understand themselves as being in control of and at home in their social world, and will only disappear when they can achieve such a self-understanding once more. It is not clear where Marx and subsequent marxists place this originating point, nor whether they really believe that pre-modern (for example, ancient) or tribal societies are significantly free of alienation. Nor is it clear just how social relations that are 'indispensable and independent of men's will' can be overcome in a community of 'associated producers', who 'subsume' reified powers under themselves and abolish the division of labour, nor what form of life such producers would lead. To these last questions, which concern the positive content of Marx's ideal of freedom, we now turn.

Communism

What, then, does emancipation promise? What are the distinctive virtues embodied in the realm of freedom? As we have seen, Marx and orthodox marxism systematically avoid explicit answers to this question as 'utopian'. Yet, as we have also seen, the marxist views of justice and injustice, exploitation and the unfreedom of alienation all presuppose an ideal of freedom. They all employ a radical critical perspective, the standpoint of 'human' society, or the realm of freedom, that cannot be adopted unless the question is answered, unless some content is given to 'communism'. In what follows, I shall focus (unless otherwise indicated) entirely upon communism's higher phase, the lower being still 'a communist society, not as it has *developed* on its own foundations, but on the contrary, just as it *emerges* from capitalist society, which is thus in every respect, economically, morally and intellectually, still stamped with the birth marks of the old society from whose womb it emerges' (Marx 1875: 23). To continue this favourite metaphor of Marx's, we are concerned here with communism full-grown and mature, unmarked by its origins—or, one might add, by whatever obstetric methods were used to bring it into being. ('Force', as Marx elsewhere remarked, being 'the midwife of every old society pregnant with a new one' [Marx 1867: 751].)

In such a society, as we have also seen, principles of justice, and more generally of *Recht*, are assumed to have withered away: they have, in Engels's words, been forgotten in practical life. By what principles or standards, then, is this society to be judged superior? What kind of morality is the 'really human morality' that it embodies? How are its claims to be validated? By an appeal to intuition? Much of Marx's and marxist writings could be seen in that light: frequent appeal is made to the reader's sense of indignation, sympathy, and also his sense of what is 'worthy of human nature'. Yet it is an elementary marxist thought that moral intuitions will be prime candidates for class-related bias: there seems to be no good marxian or marxist reason to suppose such intuitions to be universally shared, even by fully reflective agents, let alone to appeal to them in practical reasoning. Is the indicated morality deontological, then, say kantian? There is, as the neo-kantians saw, much to support this view in the canonical texts—as, for example, in Marx's condemnation of capitalist exchange relations as social relations in which each becomes a means of the other, and more generally in his frequent talk of the 'slavery', degradation, and indignity inherent in capitalist relations. Or is it, perhaps, at root utilitarian? Here too there is much support in what Marx and later marxists have said. Communism will not only be more efficient and productive; it will abolish misery, unhappiness, and frustration: as Lenin said, working people's lives would be eased and their welfare maximized. Or is it rather perfectionist, committed to an aristotelian realization of distinctively human potentialities and excellences? I have cited much evidence to support this interpretation: as, for example, Marx's claim that it is every person's 'vocation, designation, task' to 'achieve all-round development of all his abilities' (Marx and Engels 1845–6: 292).

Marx was no moral philosopher and he did not discuss the differences between these different kinds of morality. I suspect that if he had done so he would have responded that under communism they all come to one anyway: that communism will at one and the same time embody what (Marx held) is intuitively essentially 'human', respect human worth or dignity, and maximize both welfare and self-development (See Kamenka 1972). Certainly, there is within the marxist tradition no discussion of the possibility of conflicts between these various ideals. Rather than pursuing these rather recondite questions any further, I shall here simply assert that, if we are here concerned with the question 'What makes the realm of freedom really (rather than formally) free?', it is the teleological, aristotelian, perfectionist Marx we must follow (see the first epigraph to this chapter). For the freedom that capitalism denies and communism promises is systematically couched in the language of 'species powers', potentiality, and self-actualization.

In the self-transforming and self-realizing process of emancipation, one factor is crucial: free time. As Marx wrote, in the *Grundrisse*, 'free time—which is both idle time and time for higher activity—has *naturally* transformed its possessor into a different subject' (Marx 1857–8: 712). And Marx stressed this too in what is perhaps his best-known text on the subject:

In fact, the realm of freedom actually begins only when labour which is determined by necessity and mundane considerations cease; thus in the very nature of things, it lies beyond the sphere of actual material production. . . . Freedom in this field can only consist in socialised man, the associated producers rationally regulating their interchange with nature, bringing it under their common control, instead of being ruled by it as by the blind forces of Nature; and achieving this with the least expenditure of energy and under conditions most favourable to, and worthy of, their human nature. But it nonetheless still remains a realm of necessity. Beyond it begins that development of human energy which is an end in itself, the true realm of freedom, which, however, can blossom forth only with realm of necessity as its basis. The shortening of the working day is its basic prerequisite. (Marx 1861–79: iii. 799–800.)

There is, as Heller notes, a somewhat different possibility sketched out in the *Grundrisse*, where, in a visionary anticipation of automation, Marx writes of a future in which

Labour no longer appears so much to be included within the production process; rather the human being comes to relate more as watchman and regulator to the production process itself. . . . [the worker] steps to the side of the production process instead of being its chief actor. In this transformation, it is neither the direct human labour he himself performs, nor the time during which he works, but rather the appropriation of his own general productive power, his understanding of nature and his mastery over it by virtue of his presence as a social body—it is, in a word, the development of the social individual which appears as the great foundation-stone of production and of wealth. . . . As soon as labour in the direct form has ceased to be the great well-spring of wealth, labour time ceases and must cease to be its measure, and hence exchange value must cease to be the measure of use value. (Marx 1857–8: 705.)

But note that in this vision too, it is the time for free development that is crucial, only here that development takes place in the form of scientific understanding and technical control, enjoyed for their own sake as a 'vital need', within the production process itself. For, in general, Marx held that free time, *disposable time*, is wealth itself, partly for the enjoyment of the product, partly for the free activity which—unlike labour—is not dominated by *the pressure of an extraneous purpose which must be fulfilled*, and the fulfillment of which *is regarded as a natural necessity* or *a social duty*, according to one's inclinations. (Marx 1862–3: iii. 257, italics added, S.L.)

And here we come to what is, I believe, the heart of the matter. It is emancipation from the pressure of 'extraneous purposes', from 'what is regarded as a natural necessity or a social duty', that is the key to Marx's conception of 'real' freedom. This is perhaps why Marx and Engels rejected Max Stirner's view that 'in communist society there can be a question of "duties" and interests', holding these to be 'two complementary aspects of an antithesis which exists only in bourgeois society' (Marx and Engels 1845–6: 213). This, I believe, is why they thought that

Communism differs from all previous movements in that it overturns the basis of all earlier relations of production and intercourse, and for the first time treats all naturally evolved premises as the creations of hitherto existing men, strips them of their natural character and subjugates them to the power of the united individuals. . . . The reality which communism creates is precisely the true basis for rendering it impossible that anything should exist independently of individuals, insofar as reality is nevertheless only a product of the preceding intercourse of individuals. (Ibid.: 81.)

But if this is a correct interpretation, then a host of questions crowd in upon us. What makes a purpose count as 'extraneous'? Not (unless Marx is saddled with a purely subjective, phenomenological notion of alienation) whatever the agent counts as such. But what, then, are individuals' authentic, non-extraneous purposes? Marx imagines a world in which the question does not even arise, because its answer is both not in doubt and correctly understood, both obvious and true (a world, therefore, free of moral scepticism). And *are* there not natural necessities which human activity (including labour) must fulfil as a prerequisite of social co-operation in general and (as both Weber and Durkheim thought) of a complex modern social order in particular? And as for social *duties* (this takes us back to the withering away of *Recht*), is their disappearance conceivable, even in a world inconceivably more abundant and co-operative than our own—either in the sense that the required tasks now conceived as duties would disappear or that they would no longer be seen *as* duties?

Further, wider questions arise too. Marx's vision is of 'the free development of individualities . . . the general reduction of the necessary labour of society to a minimum, which then corresponds to the artistic, scientific, etc. development of the individuals in the time set free, and with the means created, for all of them' (Marx 1857–8: 706). But what does 'the free development of individualities' for all mean? That all should have an equal *opportunity* to develop whatever manifold capacities (or rather, relevant 'human' capacities) they severally have? Or that all would *in fact* realize such capacities, attaining, therefore, different levels of achievement in any one capacity and of success in developing and integrating several. Or does it mean that a maximum level of 'artistic, scientific, etc. development' will be achieved in 'society' as a whole? Whichever interpretation is the right one, it seems undeniable that it invokes a *distributive* principle, even if this is not a principle of distributive justice (in the Hume–Rawls sense), for it specifies how resources should be allocated to achieve the result in question, indicating who should get what. In fact, the three interpretations invoke three different distributive principles, viz. (1) equal (and unlimited) access to the (external) means to self-development (the maxima of such development varying from individual to individual), i.e. a version of the *equality of opportunity* principle; (2) application of whatever means (including resources) are needed to achieve universal maximal self-development, i.e. a version of the *equality of results* principle (so that all are maximizing their respective potentials); and (3) whatever distribution of resources and application of means are needed to achieve maximal artistic, scientific, intellectual, etc. development in 'society', which says nothing about equality but makes both the proper distribution and social policy depend on what that desired outcome requires.

Yet further questions arise here. First, what is to be the initial criterion of development itself, what is to count as a capacity realization, and how are specific capacity realizations to be traded off against many-sidedness and both of these against the integration of the sides into a meaningful life-plan? Second, which of the principles suggested seems most likely, as an interpretation of Marx? Does not 'the free development of individualities' mean, precisely, autonomy or *self*-development—that is, not merely the abstract possibility but the real possibility of individuals choosing *non*-development? This suggests interpretation (1). But Marx usually writes as though he favours interpretation (2)—and, almost certainly, did not distinguish between (1) and (2). But (2) poses extreme difficulties in application. It is a principle specifying who should get what, but one in which the *who* and the *what* are inter-dependent: the identity of agents is affected by some of the means in question (from special schools for the gifted to genetic engineering). If the principle is carried beyond a certain point (set by some limit to how far any given self may by altered), selves are being transformed, even created. Depending on the technical possibilities (and under communism these would surely be very great), (2) could come to mean $(2)^1$, viz. the application of whatever means are needed to achieve universal and equal maximal self-development. Sometimes, Marx and marxists have suggested that this is what is meant. But often they have suggested that they favour interpretation (3), and most probably assume that (3) would coincide with (2) and (1). But would they coincide? Take first (3) and (2). This is most unlikely, since scientific and artistic *achievement* must surely involve some succeeding where others fail. Although the latter may in some sense realize their capacities, to say that they do so as the former do is either to delude or to mock them. And perhaps some (considerable?) measure of such failure is a precondition for maximum success overall (see Elster 1985)? What then of (3) and (1)? Here perhaps distributive justice re-emerges after all, since maximal scientific and artistic development may well require the concentration of resources among the most talented, and not all resources could be so abundant for this not to be a real issue (e.g. in education and training).

Aside from these distributional questions, there are others. How is the 'free development of individualities' to be *co-ordinated*? In particular, is there not a contradiction between the image of *community*, or even of a community of communities, on the one hand, and that of society as a gigantic factory on the other? For Marx was also the author of the idea, taken up by Kautsky and Lenin, that society as a whole could be organized on the basis of the division of labour, socially controlled and regulated, rather than being subject to 'competition . . . the coercion exerted by the pressure of . . . mutual interests' (Marx 1867: 356; see Selucky 1979: 13–14). Selucky has suggested that Marx's community-concept applies to the political sphere, and his nation-wide factory concept to the economic sphere, and that the two are in conflict:

Since economic base and superstructure are to be in structural accord, one is facing a dilemma: either to accept and introduce Marx's centralised society-type organised economy and to revise Marx's concept of the decentralised community-type polity, or to accept and introduce the decentralised community-type polity and to revise Marx's concept of the centralised society-type organised economy. (Selucky 1979: 87.)

Lenin, Selucky suggests, did the former, and the Yugoslavs, in their market-based socialist economy, have sought to do the latter. Is any other solution possible?

And what of decision-making? How are the allocation of resources and the assignment of functions to be decided upon? By all, but by what methods: participatory or representative? Marx was undoubtedly committed to the ideal of direct democracy. His early conception of such democracy involved a Rousseauesque critique of the principle of representation, and the view that true democracy involves the disappearance of the state and thus the end of the separation of the state from civil society, which occurs because 'society is an organism of solitary and homogeneous interests, and the distinct "political sphere" of the "general interest" vanishes along with the distinction between governors and governed' (Colletti 1975: 44). This view reappears in Marx's writings about the Paris Commune, which he admired for its holding every delegate 'at any time revocable and bound by the formal instructions of his constituents' and 'instead of deciding once in three or six years which member of the ruling class was to misrepresent the people in Parliament, universal suffrage was to serve the people, constituted in Communes . . .' (Marx 1871a: 520). Partly because this was his view, he never addressed the procedural issue of what forms collective choice or decision-making should take under communism, whether at the lower or higher stage. And how seriously is the withering away of conflicting interests to be taken? The more seriously, the less will constitutional protections be needed, and the more Rousseauesque will communism be (whatever that might mean in a complex industrial society).

Finally, and most deeply, *who* are the individuals who are to be united under communism, and what kind of unity do they attain? Do they come with attachments, commitments, and loyalties, their identities shaped by local, regional, national, historical experiences and memories? Or are they the ultimate fruits of Enlightenment individualism: individuals who, once their class situation has withered away, are free of 'extraneous purposes', 'natural necessities', and 'social duties', with nothing existing 'independently of them'? Is the unity of communism a *community*, rooted in particularity and marked off from others? Or is it the unity of freely associating sovereign choosers, combining to attain 'the full development of human mastery over the forces of nature, those of so-called nature, as well as of humanity's own nature' (Marx 1857–8: 488)?

To none of these questions do Marx or the canonical marxist texts give answers. Indeed, as we have seen, they advance various anti-utopian arguments for *not* answering them—at least until the time comes for their solution. And this is perhaps one reason why, throughout the history of marxism in practice, and above all in power, different policies have been pursued in all these areas—the distribution of goods and resources, social policy, social and industrial organization, political and constitutional structures, nationalism and regionalism—without it being clear *which* policy was or was not compatible with the ultimate ends of communism. What the realization of these ends could mean in practice remains unclarified, indeed largely unconsidered, within marxism.

All this is serious enough. It has absolved marxists engaged in revolutionary struggles from political responsibility, and deprived those directing 'transitional' regimes of a coherent sense of direction. (To what are they transitional?) Most seriously of all, it has precluded consideration of the very relation between *Recht* and emancipation. Is emancipation conceivable, let alone feasible, without the recognition of principles of justice and of rights?

Whether it is *feasible* divides into two questions: could it exist in practice, and could it be attained, without such recognition? I have already suggested an answer to the first of these questions in the previous chapter: the circumstances of justice and the conditions of *Recht*, which require the protections and guarantees that rights afford, are not, given all that we know, ever likely to be overcome. Not only are scarcity, limited altruism, conflicting moralities, and limits upon knowledge and understanding here to stay; minorities will always need protection from the most democratic precedures, interpretations of shared goals will always differ, and there will always be a need to secure public goods and provide for future generations in a way that will not be immediately apparent and compelling to all individuals. As for the second question, and the ways in which it has been debated in the marxist tradition, it is the subject of the chapter which follows.

Is emancipation without the recognition of justice and rights *conceivable*? Joseph Raz has suggested that the co-ordinating, dispute-resolving, and damage-remedying functions of law would be needed even in 'a society of angels' (Raz 1975: 159), and presumably by the same argument principles of justice and rights would need to be recognized in such a society. On what grounds might one reject such a suggestion? Only on the ground that it takes too low a view of angels: that such angels (or rather perhaps saints) would be in such *gemeinschaftlich* or communal relations with one another that such recognition would be unnecessary. But what could such communal relations be like, and what could be their basis?

Here there seem to be only two alternatives. On the one hand, such angels or saints could agree upon and live by shared moral principles, in a kind of communist *Sittlichkeit*. They would guide what would otherwise be conflictual into harmonious and mutually advantageous behaviour, by moderating and reconciling claims on common resources, enforcing respect for others' interests and views, settling disagreements of interpretation and fact, and so on. But what could such principles be but principles of justice specifying rights?

It might, of course, be countered that these would be principles of *socialist* justice and *socialist* rights. The latter, for instance, would, it might be said, be 'more positive, less dependent on the activation of the right-holder, more directed towards the protection and furtherance of those concerns which express the needs of active and creatively productive social beings than is the case with capitalist rights', 'more organisational than political, in that they inform the co-operative social effort rather than represent demands to be disputed and traded off against each other', and 'devices to secure the benefits to be derived from harmonious communal living, not protections for the individual against the predations of others' (Campbell 1983: 213). Socialist rights, on this view,

are more typically directives and enablements than claims. They signify the proper ends and capacities of organised co-operative activity rather than the ultimate recourse of aggrieved individuals. This places them at the conscious centre of any human group organised to satisfy the needs of socialised man. (Ibid.: 148.)

But such a conception, according to which socialist *Recht* would protect and enforce certain individual interests—the needs of socialized man—directing 'the individual to goals that are at once personal and social' (ibid.: 55), already presupposes such a 'revolution in human motivation' (ibid.: 146) that a question arises whether such putatively socialist justice and socialist rights should not be otherwise described. Specifically, it appears to presuppose Pashukanis's 'social person of the future, who submerges his ego in the collective and finds the greatest satisfaction and meaning of life in this act' (Pashukanis 1924: 159–61), or rather a society of persons (angels? saints?) whose self-interests no longer conflict either with one another or with the public or collective interest. This view is essentially indistinguishable from the second answer to the question raised above: on what could the communal relations of communism be based?

That second answer is, precisely, that such relations would be between persons without any sense of a self-interest conflicting with that of others or with the public or collective interest. There is good reason to think this was indeed Marx's conception of communism: for he always tended to see self-interest as tied to civil society and private property, and as characteristic of 'egoistic man, of man separated from other men and from the community' (Marx 1843a: 162). He envisaged communism, not as the 'love-imbued opposite of selfishness' (Marx and Engels 1846: 41), but as the end of 'a cleavage . . . between the particular and the common interest' (Marx and Engels 1845–6: 47) and of the 'division of the human being into a *public man* and a *private man*' (Marx 1843a: 155); as a state in which 'man's private interest' is made to coincide with 'the interest of humanity' and 'each man' is 'given social scope for the vital manifestation of his being' (Marx and Engels 1845: 131), and in which 'the contradiction between the interest of the separate individual or the individual family and the common interest of all individuals who have intercourse with one another' has been abolished (Marx and Engels 1845–6: 46). Under such conditions, 'the individuals' consciousness of their mutual relations will, of course . . . be completely changed, and, therefore, will no more be the "principle of love" or *dévouement* than it will be egoism' (ibid.: 439).

Just what *is* the picture here? The end of 'egoism' and 'separation' between individuals and between the individual and the community, between the individual and the common interest, along with the maximal flourishing of creative self-development and individuality. On the face of it, the latter would seem to require that every individual's interests become ever more complex and that different individuals' interests become ever more diverse. The former would seem to require that these interests be directed towards and subject to the bonds uniting the members of a community.

Is there not a contradiction, or at least a tension, between the individualistic and the communitarian impulses in Marx's thought? The notion of individuality, to which as we have seen Marx was so much attached, which Georg Simmel described as 'the individualism of uniqueness (*Einzigkeit*) as against that of singleness (*Einzelheit*)', and which reached the nineteenth century through Romanticism, Goethe supplying its artistic and Schleiermacher its metaphysical foundation—this notion prescribes that 'each individual is *called* or destined to realise his own incomparable image' (Simmel 1917: 81). The notion of community, to which Marx was no less attached, which is no less rooted in the Western political tradition, pictures individuals as finding their fulfilment in reciprocity and solidarity rather than competition and self-assertion, and in mutual identification in common activities and the pursuit of common purposes. Yet Marx and Engels say that 'only within the community has each individual the means of cultivating his gifts in all directions: hence personal freedom becomes possible only within the community. . . . In the real community the individuals obtain their freedom in and through their association' (Marx and Engels 1845–6: 78). 'The unity of man with man', Marx proclaims, 'is based on the real differences between men' (Marx 1844d: 354). How is this possible?

We can imagine a range of possibilities from what we may call the minimum to the maximum picture. On the minimum picture, the diverse interests that individuals severally pursue in the course of their self-development are always overriden, when the need for choice arises, by the principle of preserving communal relations with others. The maintaining and enhancing of the latter always takes priority over individuals' self-developmental needs, wherever the two conflict. On the maximum picture, communal relations undercut rather than override conflicting interests: they enter into or help to constitute one's very conception of one's interests and one's self, and what counts as one's self-interest and one's self-development. (In the Paris Manuscripts, Marx wrote that 'activity and enjoyment, both in their content and in their *mode of existence* are *social: social* activity and *social* enjoyment' [Marx 1844c: 298].) The projects I value, the life-plans I pursue, the fulfilment I seek, and indeed my view of myself are what they are only because of the relations in which I stand to others, indeed they cannot be conceived apart from such relations. My natural inclinations and spontaneous tendencies to act, and also my considered and reflective purposes and projects, are always such as maintain and enhance the communal relations in which I stand. My inclinations are 'naturally' communal, in this sense, but should they 'unnaturally' deviate, for whatever reason, reflection will make them so. (I suppose that on the maximal maximum picture there just would be no such deviation.) On the maximum picture, it is assumed that in the 'real community', the abolition of private property, alienating social relations, and exploitation will effect a ramifying transformation upon all social relations, rendering a harmony of interests throughout social (and personal) life possible at last.

There can be no doubt that Marx inclined towards the maximum picture. But hard questions remain. Why should such a ramifying and harmonizing transformation occur? Why should it be that

with the abolition of the basis, private property, with the communistic regulation of production (and, implicit in this, the abolition of the alien attitude [*Fremdheit*] of men to their own product), the power of the relation of supply and demand is dissolved into nothing, and men once more gain control of exchange, production and the way they behave to one another. (Marx and Engels 1845–6: 48.)

Are there not other sources of interest conflict, unamenable to such structural transformation? Indeed, would not such a transformation, while abolishing the distinctive interest conflicts of capitalism, in turn generate others?

Even supposing a 'real community' to exist, what *are* its distinctive social relations? Are they face-to-face relations or do they hold between strangers; are they intimate or anonymous; are they relations of friendship, love, neighbourliness, comradeship, or kinship, or of class, ethnicity, nationality, citizenship, or common humanity; do they hold between producers, or between producers and consumers, or between citizens; are they relations of commitment and loyalty binding members to sub-communities or the community as a whole? If the society in question is of any complexity, if indeed it is a *society*, then the only possible answer is: at least all of these. But how are these various social relations themselves related? Will not the interests dictated by these various social relations be likely to conflict with one another? If so, which should have priority, and when? When, for example, should citizenship override friendship, or meeting the needs of one's family outweigh the demands of common humanity? How are we to balance the demands of consumers and producers, of locality, citizenship, and internationalism, and of all the diverse groupings—ethnic, cultural, occupational, regional, and so on— into which our social and communal attachments inevitably divide us? How can the individual, on whom all these relations bear, and who must interpret their import, avoid hard choices between their various requirements? And how could such choices be avoided in any community in which public priorities have to be decided, public choices made, and resources allocated? And how could such conflicts at the individual and at the public levels be resolved other than by appeal to deliberative moral principles, that would be principles of justice, specifying rights and obligations? Would not such principles be needed in order to define and relate those spheres to one another in a mutually acceptable manner? In short, do not even high-level, communally related, angels stand in need of *Recht*?

Plainly, the marxist tradition has not addressed such questions. Indeed, for the reasons we have discussed, it has been inhibited from subjecting even what Kropotkin decribed as the 'essential features' of its vision of emancipation to detailed scrutiny. In consequence, it is not clear whether that vision represents a development of or a decisive break with various post-Enlightenment assumptions central to Romanticism, liberalism, and democratic politics. What happens under communism to the Romantic notion of individuality, to the liberal notion of liberty as incorporating as an essential component the idea of a private space free from public interference (see Walicki 1983), and to the notion of democratic politics as the articulation of and confrontation within the public sphere of divergent views and interests? And underlying all such questions, what is the nature of the self and its interests that is to be liberated? Until such questions are faced, the marxist vision of emancipation remains an ill-defined, and thus indefensible, social and political goal.

6   Means and Ends

Point not the goal, until you plot the course,
For ends and means to man are tangled so
That different means quite different aims enforce;
Conceive the means as ends in embryo.

(Ferdinand Lasalle, *Franz von Sickingen*, quoted
in Trotsky 1938: 38 and Koestler 1940: 193.)

HUGO. I've never lied to our comrades. I . . . What use would it be
to fight for the liberation of mankind if you despised them enough
to stuff their heads with lies?

HOEDERER. I lie when I must and I despise no-one. I didn't invent the
idea of lying; it was born of a society divided into classes and each
of us inherited it at our birth. We shan't abolish lies by refusing to
lie ourselves; we must use every weapon that comes to hand to
suppress class differences.

HUGO. Not all methods are good.

HOEDERER. All methods are good when they are effective.

HUGO. Then what right have you to condemn the Regent's policy? He
declared war on the U.S.S.R. because it was the best way of
safeguarding our national independence.

HOEDERER. Do you imagine I *condemn* him? I've no time to waste.
He did what any poor fool of his caste would have done in his place.
We're not fighting men or a policy, but against the class which
produced that policy and those men.

HUGO. And the best method you can find to carry on the fight, is to
offer to share the power with them?

HOEDERER. Exactly. Today, it is the best method. [*Pause*] How
attached to your purity you are, my boy! How frightened you are
of soiling your hands, All right, stay pure! Who does it help, and
why did you come to us? Purity is an ideal for a fakir or a monk.
You intellectuals, you bourgeois anarchists, you use it as an excuse
for doing nothing. Do nothing, stay put, keep your elbows to your
sides, wear kid gloves. My hands are filthy. I've dipped them up to
the elbows in blood and slime. So what? Do you think you can govern
and keep your spirit white?

(Sartre 1948: 94–5.)

Rubashov stared through the bars of the window at the patch of blue above the machine-gun tower. Looking back over his past, it seemed now that for forty years he had been running amuck—the running amuck of pure reason. Perhaps it did not suit man to be completely freed from old bonds, from the steadying brakes of 'Thou shalt not' and 'Thou mayst not', and to be allowed to tear along straight towards the goal.

(Koestler 1940: 205.)

In the first five chapters of this book, I have tried to show how marxian and marxist theory on the one hand exposes illusions about justice and the reality of rights under capitalism and even under communism's lower phase; and on the other hand looks forward to the coming of full communism, the realm of real freedom, in which notions of justice and rights will have withered away and been forgotten in practical life. But marxists in practice, above all from the time of the October Revolution, have had other more immediate and pressing concerns: they have faced questions of tactics and strategy, loyalties and commitments, in the face of momentous world-historical events that seemed to put not only the future of socialism but also the inheritance of liberalism in jeopardy. What bearing has the theory had on how these questions have been resolved? In particular, has it established any guidelines or limits to which means are appropriate for realizing the ends envisaged? What is to be done and what is not to be done?

In examining this question, I shall focus upon three periods in the history of twentieth-century communism, when this question was debated within the marxist tradition: after the Russian Revolution when Bolshevik methods were fiercely argued over, among others by Rosa Luxemburg, Kautsky, Trotsky, and Lukács; after the Stalin purges and show trials, in the wake of the Spanish Civil War and in the face of fascism and nazism, when Trotsky debated with John Dewey and Victor Serge over means and ends; and after the Second World War, when, as Raymond Aron has put it, 'France was, *par excellence*, the battlefield on which the Cold War was fought, the war of ideologies and intellectuals' (Aron 1983a: 11), and when, among others, Sartre and Merleau-Ponty addressed the issue of dirty hands in politics and the meaning of Stalinism for those who sympathized with the communist cause.[1] In one sense, of course, the question of means and ends has pervaded the entire history of marxism, and more generally of socialism in our century: the social democratic tradition, from Revisionism to Eurocommunism, offering one kind of answer, while the leninist tradition, in its various forms, offers another. Yet, in another sense, marxism has largely avoided the question, as an issue of principle, explicit discussions of which are rare. The texts I have just cited, and will now discuss, are for that reason all the more interesting.

The question has of course become central to the writings of East Europeans—Poles, Hungarians, Czechs, and Yugoslavs—in the period, roughly between the mid-1950s and the early 1970s, when a latter-day revisionism of a 'marxist humanist' stamp seemed a real possibility (for a few it is still so), and since; and most recently it has re-emerged in response to the accumulating evidence of the Gulag Archipelago. (Now, as in the 1940s, it is a novelist who has crystallized and mobilized this response: then Koestler, now Solzhenitsyn.) But I doubt if anything new has been added in any of these discussions that is distinctively marxist when set beside the marxist discussions of the question in the earlier phases, when the issues were at their most acute, to which I now turn.

It was Engels who remarked that 'a revolution is certainly the most authoritarian thing there is': it is 'the act whereby one part of the population imposes its will upon the other part by means

[1] In his memoirs, Irving Howe writes of a

trend in Western intellectual life in the forties when, in revulsion from the brutalities of Stalinism and the rigidities of anti-Stalinist Marxism, a number of European intellectuals, especially in France, tried to plant a radical politics in the soil of humanist sentiment and ethical value. This effort produced better literary work than political guidance, since it was hard to get beyond stirring injunctions that means should be closely aligned with ends. These injunctions, fine as they were, still left us staring helplessly at the world as it was; they could at best tell us what to be or become, but very little about what to do. The Marxists, impatient with the moral frettings of writers like Camus, Silone and Chiaromonte, pointed to the necessary gap between assertions of value and necessities of practice; but the Marxists, or most of them, were blind to what was authentic in the gropings of these intellectuals. Ethical consciousness could not be a substitute for politics, yet without it politics had come to seem intolerable. (Howe 1983: 116–17.)

This trend found an echo on the American left, especially in the pages of Dwight Macdonald's journal, *Politics* (see Macdonald 1946). A similar trend occurred in Britain a decade later around the early New Left, in response to the crushing of the Hungarian uprising.

of rifles, bayonets and cannon—authoritarian means if ever there were any; and if the victorious party does not want to have fought in vain, it must maintain this rule by means of the terror which its arms inspire in the reactionaries' (Engels 1874: 639, S.L.). But is a socialist revolution any different? How 'authoritarian' must or can it be? What forms of authoritarianism can it tolerate? And under what conditions, directed against whom, and for how long? Marxists have offered interestingly different answers to these questions.

I

Rosa Luxemburg was the first to raise them, from her prison cell, in her far-seeing pamphlet on the Russian Revolution, written in 1918. On the basis of visitors' accounts and smuggled newspapers, she issued prophetic warnings about the course of the Revolution, from a position of complete, but deeply questioning, solidarity with the Bolshevik party (the 'only one in Russia which grasped the true interest of the Revolution in the first period', and 'the only party which really carried on a socialist policy' [Luxemburg 1918: 35]).

She had no illusions about the 'conditions of bitter compulsion and necessity' the Bolsheviks faced. Their experiment in proletarian dictatorship was taking place 'under the hardest conceivable conditions' (ibid.: 29, 28); and she was clear that 'a model and faultless proletarian revolution in an isolated land, exhausted by world war, strangled by imperialism, betrayed by the international proletariat, would be a miracle' (ibid.: 80). Furthermore, she was clear about the revolution's need to defend itself in order to survive:

As the entire middle class, the bourgeois and petty-bourgeois intelligentsia, boycotted the Soviet Government for months after the October Revolution and crippled the railroad, post and telegraph, and educational and adminstrative apparatus, and, in this fashion, opposed the workers' government, naturally enough all measures of pressure were exerted against it. These included the deprivation of political rights, of economic means of existence, etc., in order to break their resistance with an iron fist. It was precisely in this way that the socialist revolution expressed itself, for it cannot shrink from any use of force to secure or prevent certain measures involving the interests of the whole. (Ibid.: 65.)

What alarmed her was the elevation of such necessities into socialist norms:

The danger begins when [Lenin and his comrades] make a virtue of necessity and want to freeze into a complete theoretical system all the tactics forced upon them by these fatal circumstances, and want to recommend them to the international proletariat as a model of socialist tactics. (Ibid.: 79.)

In particular, she was alarmed by the distrust of representative democracy revealed by the dissolving of the Constituent Assembly (and the non-summoning of another); by the restriction of the right to vote to those who live by their own labour, disfranchising 'broad sections of society, many of them unable to work'; by the destruction of 'the freedom of the press, the right of association and assembly, which have been outlawed for all opponents of the Soviet regime' (ibid.: 66); by 'the means [Lenin] employs. Decree, dictatorial force of the factory overseer, draconian penalties, rule by terror' (ibid.: 79); by the 'persistent dangers of martial law' which 'leads inevitably to arbitrariness' (ibid.: 74); by 'the use of terror to so wide an extent by the Soviet government' (ibid.: 78); and, in general, by 'the tacit assumption underlying the Lenin-Trotsky theory of the dictatorship', namely, 'that the socialist transformation is something for which a ready-made formula lies completed in the pocket of the revolutionary party, which needs only to be carried out energetically in practice' (ibid.: 69).

Trotsky and Lenin had in 'the elimination of democracy' found a remedy

worse than the disease it is supposed to cure; for it stops up the very living source from which alone can come the correction of all the innate shortcomings of social institutions. That source is the active, untrammeled, energetic political life of the broadest masses of the people.

Here lies the essential basis of Rosa Luxemburg's argument. She did not defend the universal extension and maximal exercise of democracy and political freedom by reference to 'some sort of abstract scheme of justice' or 'in terms of any other bourgeois-democratic phrases', but by reference to 'the social and economic relationships' for which they are 'designed' (ibid.: 63). Her fundamental belief was twofold: that 'the only effective means in the hands of the proletarian revolution' were 'the kindling of revolutionary idealism, which can be maintained over any length of time only through the intensively active life of the masses themselves under conditions of unlimited political freedom' (ibid.: 74); and that under such conditions—above all 'general elections . . . unrestricted freedom of press and assembly . . . a free struggle of opinion'—the 'active participation of the masses' (ibid.: 77) would lead to a socialist future. They were, in short, both necessary and sufficient conditions. Socialism, as she famously wrote,

by its very nature cannot be decreed or introduced by *ukase*. It has as its prerequisite a number of measures of force—against property, etc. The negative, the tearing down, can be decreed; the building up, the positive, cannot. New territory. A thousand problems. Only experience is capable of correcting and opening new ways. Only unrestricted, effervescing life falls into a thousand new forms and improvisations, brings to light creative force, itself corrects all mistaken attempts [*sic*]. The public life of societies with limited freedom is so poverty-stricken, so miserable, so rigid, so unfruitful, precisely because, through the exclusion of democracy, it cuts off the living source of all spiritual riches and progress. . . . Socialism in life demands a complete spiritual transformation in the masses degraded by centuries of bourgeois class rule. Social instincts in place of egotistical ones, mass initiative in place of inaction, idealism which conquers all suffering, etc. (Ibid.: 72.)

In the absence of all of this, 'life dies out in every public institution, becomes a mere semblance of life, in which only the bureaucracy remains as the active element', with the eventual 'brutalisation of public life'.

There was much acumen and foresight in Luxemburg's pamphlet. These were all the more striking in coming from a comrade-in-arms, fully prepared to allow for the necessity of harsh means to achieve noble ends. As Geras has shown (Geras 1976), Rosa Luxemburg was no anarchist proclaiming that means must in all circumstances *prefigure* the ends pursued: she did not, for instance, endorse Emma Goldmann's purist view that 'no revolution can ever succeed as a factor of liberation unless the MEANS used to further it be identical in spirit and tendency with the PURPOSES to be achieved' (Goldmann 1925: 261). Yet she had a quite unrealistic set of expectations of what mass action could achieve in Russian conditions: as Peter Nettl has commented, her 'call for broad participation in Social Democratic activity was partly due to an excessive transplantation of idealised German conditions into the Russian context, just as Lenin's conditions were far too narrowly Russian to have general validity' (Nettl 1966: i. 290). There was also, on the one hand, a crucial vagueness about just which 'measures involving the interests of the whole' justify the 'use of force' (she subsequently changed her mind, for example, about the dissolving of the Constituent Assembly [see Geras 1976: 187]), and on the other, an unbounded, and ungrounded, optimism about the course and destination of revolutionary mass action, which 'corrects all mistaken attempts', in which she invested as much faith as did Lenin and Trotsky in their 'ready-made formula'.

Neither faith was shared by Karl Kautsky, whom Rosa Luxemburg called 'the official guardian of the temple of Marxism' (Luxemburg 1918: 33) and Lenin had called 'the chief of German revolutionaries' in whom 'the method of Marx' lived again (quoted Salvadori 1979: 252), but whom Lenin and Trotsky were now to treat as a 'renegade', who had betrayed both revolutionary marxism and his own past. In fact, this last charge was quite groundless, for Kautsky had always believed in the parliamentary road to socialism, in democratic freedoms and civil liberties, pluralism of parties, periodic elections, and the use of violence only as a last resort against those who refused to accept a legally constituted socialist majority government. The fact that Kautsky called all this 'the dictatorship of the proletariat' is rather a tribute to the Power of the Word in the marxist canon than to Kautsky's gift of accurate self-expression.

For Lenin in 1918 the dictatorship of the proletariat meant 'power based directly on force and unrestricted by any laws', a 'rule won and maintained by the use of violence by the proletariat against the bourgeoisie, a rule that is unconstrained by any laws' (Lenin 1918: 236). For Kautsky, this—the Bolsheviks' conception of the dictatorship of the proletariat— was quite un-marxist, 'nothing but a grandiose attempt to leap over necessary phases of development and to eliminate them by decree' (quoted Salvadori: 257), which could only issue in the dictatorship of a minority and what he came to call a 'barracks socialism' (Kautsky 1919: 207).

Kautsky's argument, well expressed in his *Terrorism and Communism*, written between 1918 and 1919 (and ever more forcefully expressed over the next decade), was that the Bolshevik regime, based on a dictatorship of party leaders, was leading away from socialism, 'transforming what should have been the social struggle for liberty, and for the raising of the whole of humanity on a higher plane, into an outbreak of bitterness and revenge, which led to the vast abuses and tortures'. He held that this was so because the Bolshevik Revolution was premature; and that socialism and 'the abolition of any form of exploitation and oppression' (ibid.: 180) to which it in turn was a means would come only under democratic and advanced economic conditions, legally enacted by a mature and organized working class: for 'wherever socialism does not appear to be possible on a democratic basis, and where the majority of the people rejects it, its time has not yet fully come' (ibid.: 220).

In Kautsky's view, the Bolsheviks had finally 'achieved the very contrary of that which they had set out to obtain. For instance, in order to come into power they threw overboard all their democratic principles. In order to keep themselves in power, they have let their socialist principles go the way of the democratic' (ibid.: 198):

Originally they were whole-hearted protagonists of a National Assembly, elected on the strength of a universal and equal vote. But they set this aside, as soon as it stood in their way. They were thorough-going opponents of the death-penalty, yet they established a bloody rule. When democracy was being abandoned in the State they became fiery upholders of democracy within the proletariat, but they are repressing this democracy more and more by means of their personal dictatorship. They abolished the piece-work system and are now reintroducing it. At the beginning of their regime they declared it to be their object to smash the bureaucratic apparatus, which represented the means of power of the old State; but they have introduced in its place a new form of bureaucratic rule. They came into power by dissolving the discipline of the army, and finally the army itself. They have created a new army, severely disciplined. They strove to reduce all classes to the same level, instead of which they have called into being a new class distinction. They have created a class which stands on a lower level than the proletariat, which latter they have raised to a privileged class; and over and above this they have caused still another class to appear, which is in receipt of large incomes and enjoys high privileges. (Ibid.: 215–16.)

In support of these claims, Kautsky pointed to the removal of all rights from the bourgeois, and the imposition on them of compulsory labour (ibid.: 169–70), to the abolition of a free press (on the 'naive assumption that there really exists an absolute truth, and that the Communists alone are in possession of that truth' [ibid.: 176]), to the 'exclusion and suppression of any kind of criticism' (ibid.: 184), and to 'the treatment of the intelligentsia', kicked 'from behind and from the front'. Further, he pointed to the role of revolutionary tribunals and secret extraordinary commissions free of all control, by means of which 'the Russian proletariat is to have pummelled into it the communist morale it lacks, in order to make it ripe for socialism' (ibid.: 182), which have 'the arbitrary power to condemn anyone who shall be denounced to them' and 'to shoot those of whom they do not approve', and are themselves corrupt (ibid.: 187); to the economic chaos within which 'such a kind of regime could only continue by having some powerful means of violence to support it, such as a blindly obedient and disciplined army' (ibid.: 204); and in general to the 'Regiment of Terror' which was 'the inevitable result of Communist methods' (ibid.: 208). Moreover, Kautsky cited Lenin, Trotsky, Bukharin, and others positively advocating such methods which he saw as deriving from 'pre-Marxist' ways of thought. The Russian Revolution had reawakened 'primitive ways of thought, and also allowed brutal and murderous forms of political and social war to come to light, forms which one had been led to believe had been overcome by the intellectual and moral rise of the proletariat' (ibid.: 157).

Accordingly, Kautsky thought,

the task of European socialism, as against Communism, is . . . to take care that the moral catastrophe resulting from a particular method of Socialism shall not lead to the catastrophe of Socialism in general; and further to endeavour to make a sharp distinction between these methods and the Marxist method, and bring this distinction to the knowledge of the masses. (Ibid.: 207.)

Kautsky's marxism thus embodied a benign, very nineteenth-century evolutionary optimism, according to which revolutions were becoming more and more humanitarian—a view for which he could cite Engels ('In proportion as the proletariat absorbs socialistic and communistic elements, will the revolution diminish in bloodshed, revenge and savagery' [Engels 1845b: 581]) and Marx himself, who, in *The Civil War in France*, praised the proletariat for having remained innocent of acts of violence.

Like Rosa Luxemburg, Kautsky saw clearly into the dark side of Bolshevik practice and foresaw (and later observed) its grim aftermath; and he joined her in seeing their means as subverting their ends. ('One should just as little strive to defend one's principles by surrendering them, as to defend one's life by sacrificing what gives that life content and purpose' [Kautsky 1919: 210].) Unlike her, he had little to say about the fearful dangers and pressures the Bolsheviks faced; since he believed their whole enterprise premature, he had no interest in the question: which of their measures were unavoidable? But like her he had faith in the availability of a relatively clean means of realizing socialism. For her, it was mass revolutionary action; for him it was the open parliamentary road that, in the fullness of time, would lead the advanced European working class to its goal of emancipation. It was for this reason that *neither* of Bolshevism's most acute and prophetic marxist critics really faced the issue of means and ends: they each assumed that the revolutionary transition, properly conducted, would solve the problem.

The inhibitions about dirty hands expressed by Rosa Luxemburg and Kautsky were of course felt in Russia and within the early Bolshevik leadership itself. Consider, for example, the issue of torture. Roy Medvedev tells the following story as 'typical of the young Soviet State'. In the summer of 1918 the Cheka uncovered a conspiracy against the Soviet regime headed by the British diplomat Bruce Lockhart. The conspirators were arrested and Lockhart expelled from the Soviet Union. A new Moscow journal, the *Cheka Weekly*, published a letter under the heading 'Why the Kid Gloves?' from the Chairman of the Nolinsk Party Committee and Cheka, asking:

. . . why did you not subject him, this Lockhart, to the most refined tortures, to get information and addresses. . . . With these measures you could easily have discovered a whole series of counter-revolutionary organisations, perhaps even have eliminated the possibility of future financing . . . do you think that subjecting a man to horrible tortures is more inhumane than blowing up bridges and warehouses of food with the purpose of finding an ally, in the torments of starvation, for the overthrow of the Soviet regime?

The letter, it seems, caused widespread indignation in Party circles. Readers sent letters of protest to newspapers, and some were published. Sverdlov's 'indignation was boundless', and the question was raised at the Praesidium, the highest governmental body, which resolved that the thoughts expressed in the article

are in gross contradiction with the policy and tasks of the Soviet regime. Although the Soviet regime resorts of necessity to the most drastic measures of conflict with the counterrevolutionary movement, and remembers that the counterrevolution has taken the form of open armed conflict, in which the proletariat and poor peasants cannot renounce the use of terror, the Soviet regime fundamentally rejects the measures advocated in the indicated article, as despicable, dangerous and contrary to the interests of the struggle for Communism.

They closed down the *Cheka Weekly*, dismissed the article's authors from their jobs, and forbade them from holding office in the Soviet government (Medvedev 1968: 262).

Despite this early rejection of torture (things were to be very different in the 30s and 40s), the measures and attitudes that so alarmed Rosa Luxemburg and Kautsky were real enough during the Red Terror, especially in so far as the Cheka was concerned: in the name of combating counter-revolutionary subversion, it was actively engaged in the summary arrest, execution, or imprisonment of all those it deemed to be class or party enemies. To quote an official document of the time, 'In its activity the Cheka is completely independent, carrying out searchings, arrests, shootings, afterwards making a report to the Council of People's Commissars and the Soviet Central Executive Committee' (cited Chamberlin 1935: ii. 79). As Victor Serge later wrote, the Cheka was given 'the right to apply the death penalty on a mass scale and in secret, without hearing the accused, who were unable to defend themselves and whom in most cases their judges did not even see' (Serge 1939a: 5). The Cheka's full title was 'Extraordinary Commission for the Repression of Counter-revolution, Sabotage, Speculation and Desertion'. Serge's later comment was:

If the necessity for secret procedures could reasonably be invoked in the case of conspiracy, is it proper to invoke it for the housewife who sells a pound of sugar that she has bought (speculation), the electrician whose fuses blow (sabotage), the poor lad who gets fed up with the front line and takes a trip to the rear (desertion), the socialist or the anarchist who has passed some remark or other in the street, or had some comrades together at home (agitation and illegal assembly)? Cases of this sort literally swamped those of conspiracy, whether genuine or non-existent . . . in all the different kinds of case that it dealt with, the Cheka made a frightful abuse of the death penalty. (Ibid.)

Was all this necessary? Serge later argued that it was not:

During the civil war there was perfect order behind the front itself, in the interior of Soviet territory. Travellers in these parts have described this plainly enough. There was nothing to prevent the functioning of regular courts, which might in certain cases have sat *in camera*, before which the accused would have been able to defend themselves. Would not the revolution have enhanced its own popularity by unmasking its true enemies for all to see? And correspondingly, the abuses which arose inevitably from the darkness would have been avoided. (Ibid.)

The Soviet leaders (and indeed Serge himself at the time) were not reluctant to defend such measures and activities, above all against Kautsky: terror was needed to defend the Revolution. And there is, of course, a case for the view that

no government could have survived in Russia in those years without the use of terrorism. . . . The national morale was completely shattered by the World War. No one, except under extreme compulsion, was willing to perform any state obligation. The older order had simply crumbled away; a new order, with new habits and standards of conduct, had not yet formed; very often the only way in which a governmental representative, whether he was a Bolshevik commissar or a White officer, could get his orders obeyed was by flourishing a revolver. (Chamberlin 1935: ii. 81.)

But the Bolshevik self-justifications, like the measures they permitted and encouraged, went very far beyond what this minimal case could allow. Ironically enough, in view of their subsequent fates, it was Bukharin and Trotsky who wrote most intransigently in defence of such measures.

Trotsky's own *Terrorism and Communism*, published in 1920 specifically in reply to Kautsky's, was a classic of this genre. Trotsky saw Kautsky's 'tearful pamphlet' as directed 'against revolutionary resoluteness' (Trotsky 1920: 33). He scorned Kautsky's 'fetishism of the parliamentary majority' (ibid.: 45) and his belief in democracy as 'a clear and solitary path to salvation' (ibid.: 51), while acknowledging 'certain agitational and political advantages of . . . a "legalised" transition to the new regime' (ibid.: 64; 'Hence our attempt to call a Constituent Assembly,' ibid.: 104). But 'Kautsky's ship was built for lakes and quiet harbours, not at all for the open sea, and not for a period of storms.' In support, Trotsky cited Marx, for whom 'all the questions of the proletarian state' were decided 'according to the revolutionary dynamics of living forces, and not according to the play of shadows upon the market-place screen of parliamentarism'. History, he wrote,

has not transformed the nation into a debating society solemnly voting the transition to the social revolution by a majority of votes. On the contrary, the violent revolution has become a necessity precisely because the imminent requirements of history are helpless to find a road through the apparatus of parliamentary democracy. (Ibid.: 58.)

'The Pharisees of democracy', wrote Trotsky, 'speak with indignation of the repressive measures of the Soviet Government, of the closing of newspapers, of arrests and shooting.' But 'our problem' is 'to throttle the class lie of the bourgeoisie and to achieve the class truth of the proletariat' (ibid.: 180):

We are fighting a life-and-death struggle. The press is a weapon not of an abstract society, but of two irreconcilable, armed and contending sides. We are destroying the press of the counter-revolution, just as we destroyed its fortified positions, its stores, its communications and its intelligence system. (Ibid.: 80.)

As for the Extraordinary Commissions, they 'shoot landlords, capitalists and landlords who are striving to restore the capitalist order' (ibid.: 78–9). As for terror, it 'can be very effective against a revolutionary class which does not want to leave the scene of operations. *Intimidation* is a powerful weapon of policy, both internationally and internally. War, like revolution, is founded upon intimidation' (ibid.: 78). The 'Red Terror is a weapon utilised against a class, doomed to destruction, which does not wish to perish'; without it, 'the Russian bourgeoisie, together with the world bourgeoisie, would throttle us long before the coming of the revolution in Europe' (ibid.: 83). Such terror,

the State terror of a revolutionary class can be condemned 'morally'
only by a man who, as a principle, rejects (in words) every form of
violence whatever—consequently every war and every rising. For this
one has to be merely and simply a hypocritical Quaker. (Ibid.: 78.)

Trotsky's position was clear-cut:

Who aims at the end cannot reject the means. The struggle must be
carried on with such intensity as actually to guarantee the supremacy
of the proletariat. If the socialist revolution requires dictatorship—'the
sole form in which the proletariat can achieve control of the State' [the
quotation is from Kautsky]—it follows that the dictatorship must be
guaranteed at all cost. (Ibid.: 46.)

The end was emancipation, in the canonical marxist version:

The root problem of the party, at all periods of its struggle, was to create
the conditions for real, economic, living equality for mankind as a
member of a unified moral commonwealth. (Ibid.: 62.)

Under socialism,

there will not exist the apparatus of compulsion itself, namely, the
state: for it will have melted away entirely into a producing and
consuming commune. None the less, the road to socialism lies through
a period of the highest possible intensification of the principle of the
state. . . . Just as a lamp, before going out, shoots up in a brilliant flame,
so the state, before disappearing, assumes the form of the dictatorship
of the proletariat, *i.e.* . . . . the most ruthless form of state, which
embraces the life of the citizens authoritatively in every direction. (Ibid.:
177.)

The revolution required 'of the revolutionary class that it should
attain its end by all methods at its disposal—if necessary by an
armed rising: if required, by terrorism' (ibid.: 77). If repression
is required, 'The question of the form of repression, or of its
degree, of course, is not one of ''principle''. It is a question of
expediency' (ibid.: 78). And so Trotsky totally rejected the
criticisms of 'the high priests of Liberalism and Kautskyianism'
(ibid.); it would , he claimed, 'not be difficult to show, day by
day through the history of the civil war, that all the severe
measures of the Soviet government were forced upon it as
measures of revolutionary self-defence' (ibid.: 112).
   Trotsky's case for the defence did not directly address

Kautsky's charge that what was being defended was not and could not be a socialist revolution, since it was in every sense premature. His counter-charge was that no proletarian revolution, however mature, could avoid ruthlessness and violence, that Kautsky's parliamentary road was a mirage. As for the question of which means were appropriate and permissible, of the 'form of repression' and its 'degree', his answer here could not be clearer: it was a matter of expediency, not principle.

It was left to Gyorgy Lukács to define such expediency in relation to ethical principles in his 'Tactics and Ethics', written in the months preceding the short-lived Hungarian Soviet Republic in 1919 (in which he became Commissar of Education and took a leading role in cultural affairs). If, as Lukács later wrote, Kautsky 'provided the theoretical underpinning for precisely what is *central* to Bernstein's conception of history, namely the notion of *peaceful evolutionary progression towards socialism*' (Lukács 1924: 128), it was Lukács who provided such an underpinning, on a high philosophical plane, or at least in inflated philosophical terms, for Lenin's and Trotsky's conceptions, and for Bolshevik methods. (I shall not here discuss the question of his later ambiguous attitude towards Stalinism.)

'Every essentially revolutionary objective', according to Lukács, 'denies the moral raison d'être and the historico-philosophical appositeness of both present and past legal orders'; and 'the tactics of the revolutionary classes and parties' (the 'means by which [they] achieve their declared aims') are determined by how best to achieve it. How can this be known? Lukács's answer was to say, with Marx, that the 'ultimate objective' was not an 'ideal' abstracted from and imposed on reality, but a '*reality which has to be achieved*', a transcendent objective become immanent, a goal which becomes realized through the class struggle, as it is increasingly understood and pursued by the class-conscious proletariat (Lukács 1919a: 3–4). All this occurs according to 'the logic of history' (ibid.: 5), which governs a 'world-historical process which leads through these class interests and class struggles to the final goal: *the classless society and the liberation from every form of economic dependence*' (ibid.: 17), a society that 'no longer knows either oppressors or oppressed', in which 'the blind power of economic forces must, as Marx says, be broken and replaced by a higher power which corresponds more exactly to the dignity of man' (ibid.: 5). Class consciousness provided 'a yardstick for the correctness of *immediate tactics*' (ibid.: 17), but to know what tactics were required by the 'ultimate objective' at '*moments of world crisis*' (such as that currently existing) it was necessary to be '*conscious of the world-historical mission of the proletariat's class-struggle*' (ibid.). It was

this consciousness which makes Lenin the leader of the proletarian revolution. This consciousness—in Hegelian terms, the development towards self-consciousness of society, the self-discovery of the Spirit seeking itself in the course of history—the consciousness which recognises the world-historical mission; this consciousness alone is cut out to become the intellectual leader of society. (Ibid.: 17–18.)

In short, the 'decisive criterion of socialist tactics' was 'the philosophy of history' (ibid.: 5), which dictated a certain understanding of 'the *meaning* of the class struggle of the proletariat' (entailing, among other things, the acceptance of Lenin's leadership). Once accepted as true, such an understanding enabled one to know where history was going and what the situation required, and to act in an ethical way, that is conscientiously and responsibly.

In the light of this understanding, all means could be judged:

all means by which the historico-philosophical process is raised to the conscious and real level are to be considered valid, whereas all means which mystify such consciousness—as for example acceptance of the legal order, of the continuity of historical development, let alone the *momentary* material interests of the proletariat—are to be rejected. (Ibid.: 6.)

Concretely, this meant that 'every gesture of solidarity with the existing order is fraught with . . . the danger that the feeling of solidarity will take root in *that* form of consciousness which necessarily obscures the world-historical consciousness, the awakening of humanity to self-consciousness' which understands the class struggle as a 'means whereby humanity liberates itself, a means to the true beginning of *human* history' (ibid.).

Lukács's argument was that the appropriate means to that end—the 'tactically correct [*sic*] collective action' (ibid.: 7)—

lay in the systematic promotion of proletarian class struggle, guided by a 'sense of world history' (though he did not believe in the kind of human science that could predict and thus guide tactics with certainty: it could 'indicate only possibilities'; ibid.: 10). The relation between tactics and ethics lay in his claim that the ethically conscientious and responsible individual will be guided by that sense and will follow the tactics it determines. Practically, that meant, at the political level, that 'for every socialist . . . morally correct action is related fundamentally to the correct perception of the given historico-philosophical situation, which in turn is only feasible through the efforts of every individual to make this self-consciousness conscious for himself' (ibid.: 9). But had 'the historical moment already arrived which leads—or rather leaps—from that of steady approach to that of the realisation?' (ibid.). Lukács, in early 1919, seemed not to doubt that it had, or that it would be seized.

And yet he gave no answer to the question we are here addressing: which tactics are acceptable and which impermissible? He merely stated that the means that would 'bring the goal nearer to self-realisation' (ibid.: 5), indicating the 'path' to 'the salvation of society', was the development of class consciousness and 'the sense of world history' (ibid.: 16), and that would in turn dictate 'tactically correct collective action' (ibid.: 7) and 'morally correct action' (ibid.: 9) (though he allowed that this might involve the individual in a tragic conflict, incurring guilt by following an 'imperative of the world-historical situation, a historico-philosophical mission' [ibid.: 10]). But given the inevitability of disagreements about what that consciousness and sense entailed in respect both of ends and of means—about what, in short, *is* 'correct'—there needed to be some authoritative interpretation, and thus interpreter. It is hard to think of a doctrine better tailored to justify the guiding role of the vanguard party. Whereas Trotsky, Lenin, and the rest took the end as given, and resolutely embraced whatever means they took to be necessary to achieve it, Lukács sanctified the end (with his talk of social 'salvation' and the proletariat's 'mission') and both the tactical and the ethical 'correctness' of the means the Bolsheviks saw as necessary.

II

Trotsky returned to the theme of means and ends in his pamphlet *Their Morals and Ours* (Trotsky 1938), written in Mexico in February 1938, which evoked a reply later that year entitled 'Means and Ends' from John Dewey (Dewey 1938) (who had served as Chairman of the International Commission of Inquiry into the Moscow Trials which had just vindicated Trotsky of the crimes imputed to him by the prosecution and the forced confessions of the defendants). Their exchange is remarkable, not least for the fierce honesty and clarity of Trotsky's argument, and the precision with which Dewey exposed one of its central assumptions.

Does 'the end justify the means'? In clarifying his position, Trotsky showed one, but only one, of the ways in which this often-repeated formula is crucially ambiguous. The Jesuit order, he maintained, never taught

that *any* means, even though it be criminal from the point of view of Catholic morals, was permissible if it only led to the 'end', that is, to the triumph of Catholicism. Such an internally contradictory and psychologically absurd doctrine was maliciously attributed to the Jesuits by their Protestant and partly Catholic opponents who were not shy in choosing the means for *their* own ends.

What the Jesuits actually taught was

that the means itself can be a matter of indifference but that the moral justification or condemnation of the given means flows from the end. Thus shooting in itself is a matter of indifference; shooting a mad dog that threatens a child—a virtue; shooting with the aim of violation or murder—a crime. (Trotsky 1938: 12–13.)

The end, in short, makes a difference to how the means is to be judged. (But, we must ask, *what* difference? Trotsky, in his strikingly friendly interpretation of the Jesuitical doctrine, clearly in effect answered: *all* the difference. The means have no moral weight and do not enter into the moral scales; only the purpose counts. Trotsky's example looks more telling than it is, just because it does not bring clearly into view what immoralities and illegalities this doctrine can license. Replace 'mad dog' by 'mad man' and this becomes clear. And does a noble purpose make every action aimed at its realization and likely to realize it, however horrible or ruthless, right, even virtuous?) Trotsky gave other examples of the difference that purpose and context could make to the assessment of means:

Under 'normal' conditions a 'normal' man observes the commandment: Thou shalt not kill! But if he kills under exceptional conditions for self-defence, the jury acquits him. If he falls victim to a murderer, the court will kill the murderer. . . . The most 'humane' governments, which in peaceful times 'detest' war, proclaim during war that the highest duty of their armies is the extermination of the greatest possible number of people. . . . History has different yardsticks for the cruelty of the Northerners and the cruelty of the Southerners in the Civil War. A slave-owner who through cunning and violence shackles a slave in chains, and a slave who through cunning and violence breaks the chains—let not the contemptible eunuchs tell us that they are equal before a court of morality! . . . The life and death struggle is unthinkable without military craftiness, in other words, without lying and deceit. May the German proletariat then not deceive Hitler's police? (Ibid.: 16, 29, 33.)

From all of which Trotsky concluded that one should not 'apply self-same moral norms to the oppressors and the oppressed' (ibid.: 31), and further that the appeal to such norms is 'not a disinterested philosophical mistake but a necessary element in the mechanics of class deception' (ibid.: 16).

But what, then, were the ends of socialism, and what means did *they* justify? To the first question, Trotsky answered: 'increasing the power of man over nature and the abolition of the power of man over man' (ibid: 36). To the second, he responded by raising and seeking to answer a series of questions that critics (mostly on the left) had raised against him, and, indeed, against 'marxist amoralism' in general. What of 'lying, violence and murder'? Are these not incompatible with 'a healthy socialist movement'? (ibid.: 27). To these questions, he replied with a further question:

What, however, is our relation to revolution? Civil war is the most severe of all forms of war. It is unthinkable not only without violence against tertiary figures but, under contemporary technique, without killing old men, old women and children. (Ibid.)

And, recalling the recent experience of the Spanish Civil War, he asserted that 'Whoever accepts the end: victory over Franco, must accept the means: civil war, with its wake of horrors and crimes.' From which he concluded that 'the *end* (democracy or socialism) justifies, under certain conditions, such *means* as violence or murder. Not to speak about *lies*!' (ibid.).

But did not 'lying and violence "in themselves" warrant condemnation?'

Of course, even as does the class society which generates them. A society without social contradictions will naturally be a society without lies and violence. However, there is no way of building a bridge to that society save by revolutionary, that is, violent means. The revolution itself is a product of class society and of necessity bears its traits. (Ibid.)

But was not civil war a 'sad exception': in 'peaceful times', surely a 'healthy socialist movement should manage without violence and lying'. Such an answer, he wrote,

represents nothing but a pathetic evasion. There is no impervious demarcation between 'peaceful' class struggle and revolution. Every strike embodies in an unexpanded form all the elements of civil war. . . . Thus 'lying and worse' are an inseparable part of the class struggle even in its most elementary forms. (Ibid.: 28.)

But, Trotsky relentlessly asked, does that mean that 'in achieving this end anything is permissible?' His answer is highly instructive (and we shall discuss it further):

That is permissible, we answer, which *really* leads to the liberation of mankind. Since this end can be achieved only through revolution, the liberating morality of the proletariat of necessity is endowed with a revolutionary character. It irreconcilably counteracts not only religious dogma but all kinds of idealistic fetishes, those philosophic gendarmes of the ruling class. It deduces a rule for conduct from the laws of development of society, thus primarily from the class struggle, this law of all laws. (Ibid.: 37.)

But 'does it mean that in the class struggle against capitalists all means are permissible: lying, frame-up, betrayal, murder, and so on?' Trotsky's answer was this:

Permissible and obligatory are those and only those means . . . which unite the revolutionary proletariat, fill their hearts with irreconcilable hostility to oppression, teach them contempt for official morality and its democratic echoers, imbue them with consciousness of their own historic mission, raise their courage and spirit of self-sacrifice in the struggle. Precisely from this it flows that *not* all means are permissible.

When we say that the end justifies the means, then for us the conclusion follows that the great revolutionary end spurns those base
means and ways which set one part of the working class against other
parts, or attempt to make the masses happy without their participation;
or lower the faith of the masses in themselves and their organisation,
replacing it by worship for the 'leaders'. Primarily and irreconcilably,
revolutionary morality rejects servility in relation to the bourgeoisie
and haughtiness in relation to the toilers, that is, those characteristics
in which petty-bourgeois pedants and moralists are thoroughly steeped.

But what, finally, is the answer to the question: 'what is
permissible and what is not permissible in each separate case'?
Trotsky's view was that there could be no 'automatic answers':

Problems of revolutionary morality are fused with problems of
revolutionary strategy. The living experience of the movement under
the clarification of theory provides the correct answer to these problems.
. . . The end flows naturally from the historical movement. Organically
the means are subordinated to the end. (Ibid.)

Underlying all of this, a number of key assumptions should
be noted. First, (contra Rosa Luxemburg) that the revolution,
rather than 'prefiguring' the future, 'is a product of class society
and of necessity bears its traits': thus its lying and violence are
'generated' by the pre-existing class society and 'an inseparable
part of the class struggle'. Second (contra Kautsky), that there
is no path to emancipation, no way of 'building a bridge' to 'a
society without social contradictions [and] lies and violence'
save by 'revolutionary, that is, violent means'. And third, (with
Lukács) that problems of revolutionary morality, strategy, and
tactics are one; and that 'the living experience of the movement
under the clarification of theory provides the correct [sic]
answer' to them.

Dewey, in his short reply to Trotsky, unearthed a further, deep
and crucial feature of the argument: that Trotsky (and more
generally orthodox marxists) did not subject means and ends to
the kind of examination that could warrant any rational
conclusion as to whether the means are justified or not. Dewey
shared Trotsky's consequentialism, holding that 'the end in the
sense of consequences provides the only basis for moral ideas
and action, and therefore provides the only justification that can
be found for the means employed' (Dewey 1938: 52). But he
pointed to an ambiguity (ignored by Trotsky) in the notion of

an 'end': it could either refer to an 'end-in-view' or to actual objective consequences. The maxim that 'the end justifies the means' had a bad name just because, if 'end' is interpreted in the former sense only, any act could be justified provided that the actor is sincere. The 'real question', according to Dewey, was 'not one of present belief but of the objective grounds upon which it is held: namely the consequences that will actually be produced by [the chosen means]' (ibid.: 53). One would therefore expect that

with the idea of the liberation of mankind as the end-in-view, there would be an examination of *all* means that are likely to attain this end without any fixed preconceptions as to what they *must* be, and that every suggested means would be weighed and judged on the express ground of the consequences it is likely to produce. (Ibid.)

However, Dewey noticed that, on the contrary, 'means are "deduced" from an independent source, an alleged law of history which is *the* law of all laws of social development' (ibid.: 54):

Since the class struggle is regarded as the *only* means that will reach the end, and since the view that it is the only means is reached deductively and not by an inductive examination of the means-consequences in their interdependence, the means, the class struggle, does not need to be critically examined with regard to its actual objective consequences.

And so, Dewey concluded, 'we are back in the position that the end-in-view (as distinct from objective consequences) justifies any means in line with the class struggle and that it justifies the neglect of all other means.' In short, the whole case was prejudged; it was as though a biologist were 'to assert that a certain law of biology which he accepts is so related to the end of health that the means of aiming at health—the only means—can be deduced from it, so that no further examination of biological phenomena is needed'.

Furthermore, even if the class struggle were, after investigation, shown to be the best means, how should it be carried on? To answer this too, a 'free and unprejudiced' examination of alternatives was required, instead of which

The belief that a law of history determines the particular way in which the struggle is to be carried certainly seems to tend toward a fanatical and even mystical devotion to certain ways of conducting the class struggle to the exclusion of other ways of conducting it. (Ibid.: 55.)

So it was, according to Dewey, that 'orthodox Marxists' (such as Trotsky) transferred their 'allegiance from the ideals of socialism' and a rational assessment of the means of attaining them, to 'the class struggle as the law of historical change', which 'makes all moral questions, that is, all questions of the end to be finally attained, meaningless' (ibid.: 56). After hearing and questioning Trotsky for nearly two weeks in the latter's fortified villa in Mexico City, Dewey concluded: 'It was tragic to see such brilliant intelligence locked up in absolutes' (Dewey *et al.* 1969: 432–3). By this Dewey meant what he saw as marxism's illusory certainties, as exemplified by Trotsky: certainty that its end-in-view was indeed in prospect, and certainty about which means were required, and thus permitted, to bring it into being.

Dewey's objection—that Trotsky prejudged the 'real question', assuming rather than investigating the claim that Bolshevik means would lead to socialist ends—was stated in another way by Victor Serge, moral leader of the French intellectual opposition, who had in earlier years justified Bolshevik practice in remarkably élitist terms, redolent of anarchosyndicalism rather than marxism, and held strong views on the necessity of terror in revolutions threatened by death, seeing the Red Terror as the only possible alternative to a White Terror. In 1936, having spent eighty-five days in a GPU cell, Serge denounced Stalinism, proclaiming 'it is untrue, a hundred times untrue that the end justifies the means' (cited Caute 1964: 135). There were, Serge wrote, three essential points which 'take precedence before all tactical considerations', and on these he remained 'though it may cost me dear, an open and dedicated intransigent, who will have to be coerced to keep quiet':

I. *Defence of man. Respect for man.* Man must be given his rights, his security, his value. Without these, there is no Socialism. Without these, all is false, bankrupt and spoiled. I mean: man whoever he is, be he the meanest of men—'class-enemy', son or grandson of a bourgeois, I do not care. It must never be forgotten that a human being is a human being. Every day, everywhere, before my very eyes this is being forgotten and it is the most revolting and anti-Socialist thing that could happen. . . .

II. *Defence of the truth.* Man and the masses have a right to the truth.

I will not consent either to the systematic falsification of history or to the suppression of all serious news from the Press (which is confined to a purely agitational role). I hold truth to be a precondition of intellectual and moral health. To speak of truth is to speak of honesty. Both are the right of men.

III. *Defence of thought*. No real intellectual inquiry is permitted in any sphere. Everything is reduced to a casuistry nourished on quotations. . . . I hold that Socialism cannot develop in the intellectual sense except by the rivalry, scrutiny and struggle of ideas; that we should fear not error, which is mended in time by life itself, but rather stagnation and reaction; that respect for man implies his right to know everything and his freedom to think. (Cited Serge 1942–3: 282–3.)

Trotsky bitterly attacked his old ally and admirer Serge (in *Their Morals and Ours* and in *The Moralists and Sycophants against Marxism* in 1939), not least for presuming to suggest that the seeds of Stalinism may have lain in Leninism and Bolshevism itself. Serge, commenting on Trotsky's pamphlet, saw in its 'incisive, clear and merciless style' the 'dominant tone of the Bolshevik word in the classic years' and 'an echo of the imperious, intractable style of Karl Marx, the polemicist'. It implied, Serge commented, 'in every line, in every word . . . a pretension to the monopoly of the truth, or—more accurately—an absolute conviction of the possession of the truth' (Serge 1939b).

Serge, however, admired Trotsky's denunciation of bourgeois moralism:

Trotsky is right, superbly right when he attacks the hypocrisy of conventional morals, the bourgeois class-spirit of the churches, universities and intellectual cliques; when he tracks the mediocrity of liberal idealism down to its last hiding-place; when he argues that human sentiment easily takes a subordinate place in the class struggle; when he legitimates the severities of civil war. Here, to one's great reassurance, the voice of the Russian Revolution's daring militants sounds once again. (Ibid.)

On the other hand, Trotsky neglected certain questions which touch the very nature of socialism, 'whether conceived as goal or as activity':

A Socialist revolution is not made and will not be made by scrambling in the mud—for all the old weapons of reaction. There are some kinds of action in civil war, in government, in debate, in organization which may be in certain ways efficient and sometimes even the easiest solution but which revolutionaries and Socialists must rigorously forbid to themselves on pain of ceasing to be revolutionaries and Socialists. All the old methods of the social struggle are not good for us because all of them do not lead equally to the end we pursue. We shall prove the strongest if we attain a higher degree of consciousness than our adversaries: if we are superior in firmness, vision, energy and humanity. These four terms are inseparable in reality; they form a single whole. (Ibid.)

In particular, one had to ask whether

the Bolshevik regime itself did not have certain weaknesses, certain inherent or functional vices which could facilitate the path of the bureaucratic usurpers. There is no escape from the posing of this question. Trotsky does not want to pose it. He proceeds from the notion of an ideal Bolshevism, without reproach or fault, whose history up to 1923 . . . remains blameless and unassailable. (Ibid.)

III

Trotsky's pamphlet evoked a response strikingly different from both Dewey's and Serge's in Jean-Paul Sartre who, writing in post-Liberation France, was himself preoccupied with the problem of means and ends and dirty hands in politics, which he confronted directly in *Les Mains sales* (1948) and *Le Diable et le bon Dieu* (1951). In the latter play, Sartre has the violent revolutionary peasant leader say to the pacifist Tolstoyan Goetz: 'In a single day of virtue you have created more deaths than in thirty five years of malice', and Goetz reflects: 'On this earth at present Good and Evil are inseparable. I agree to be bad in order to become good' (Sartre 1951: 224). Sartre certainly appreciated Trotsky's clear-sighted *engagement*: as he was to write in criticism of the personal and humanitarian morality of Albert Camus, with its stress on individual refusal and limited objectives clearly envisaged:

To merit the right to influence men who are struggling, one must first participate in their struggle, and this first means accepting many things, if you hope to change a few of them. . . . It is not a question of whether history has a meaning or whether we should deign to participate in it, but to try, from the moment we are in it up to the eyebrows, to give history that meaning which seems best to us, by not refusing our participation, however weak, to any concrete action which may require it. (Cited Caute 1973: 351.)

Yet Sartre's reaction to *Their Morals and Ours* focuses subtly on its convergences with, as well as its divergences from, bourgeois morality. Trotsky's 'fine but short' book, he noticed, invoked bourgeois criteria (Sartre 1947–8: 167). According to Trotsky, Lenin remained 'faithful to one and the same ideal throughout his whole life', devoting 'his whole being to the cause of the oppressed', displaying 'the highest conscientiousness in the sphere of ideas and the highest fearlessness in the sphere of action', and maintaining 'an attitude untainted by the least superiority to an "ordinary" worker, to a defenceless woman, to a child'. Bourgeois moralists could only applaud such virtues (though they differed about the nature of oppression and how to remedy it). Similarly, Trotsky wrote of the Bolshevik Party that

in the period of its revolutionary ascendance, that is, when it actually represented the proletarian vanguard, it was the most honest party in history. Wherever it could, it, of course, deceived the class enemies; on the other hand it told the toilers the truth, the whole truth and nothing but the truth.

Bourgeois moralists too, while condoning strategems of war, condemned lying to one's friends—though they differed about who one's friends and enemies were (ibid.: 168).

Furthermore, Sartre observed that Trotsky's 'ultimate end'— 'social antagonisms suppressed, man becomes an end for man, lying and violence are banished, all the powers of the human species are directed towards the conquest of nature'—was 'a Kantian ideal: it is the Kingdom of Ends'. The class struggle by itself justified nothing: it was 'necessary to be *on the side of the oppressed*', since:

1. oppression deserves condemnation. Why, if not in the name of human morality? 2. precisely because the sole human group in a position to conceive of a human morality is the oppressed who pose as end a society without classes and therefore without violence, without lying, for a liberated man.

But the bourgeois also aspired to a classless society. Yet 'he aims to realise that society *at once* by a simple interior disposition:

between a worker and an employer, both animated by generous ideas, the class barrier disappears, and only *men* remain' (ibid.: 169).

What, then, of means? Which means did the absolute end require? Trotsky's answer, as we have seen, was: those that really lead to and are not really destructive of it. Sartre observed that, unlike that of the bourgeois, Trotsky's end 'can be realised and once realised is self-perpetuating. . . . a classless society once realised, it is a permanent fact that one does not lie, because there are no longer any *motives* for lying'; and 'it needs an organisation to realise it in History, which, far from being eternal and transcendent to History, is realised in and through it, and within a limited future.' Nevertheless, the end 'functions as an ideal', a principle regulating adopted means. Of course, if 'the science of history exists', linking means and ends deterministically, then 'the normative character of the end is significantly attenuated'. But 'if *judgement* (in the kantian sense) retains a place—that is the hypothesis of a future and the invention of a way of acting—then the regulative and normative end intervenes as an imperative' (ibid.: 169–70).

What, then, for Sartre, was the difference between the Trotskyian position and that of bourgeois and democratic morality?

Both conceive a Kingdom of Ends. But the second conceives of it as being capable of being realised atemporally through the pure accord of good wills and, in the end, recognises that this realisation, though always possible, is never in fact achieved. Hence its pessimism allied to an *a priori* optimism. Trotskyism, by contrast, begins from the non-existence of the fact of this Kingdom of Ends and from the historical impossibility of realising it at any given moment. In other words, it is not original sin, or human wickedness, or the separation of consciences that it blames for preventing that realisation, but rather a precise historical situation, namely the oppression of man by man.

However, the Trotskyist position admitted

1. that the Kingdom of Ends will maintain itself spontaneously once the liquidation of classes has been accomplished; 2. that the very situation of the oppressed contains a dynamic element which will allow that Kingdom of Ends to be realised. In a word, that there is, if not in the 'human nature' of bourgeois moralists, at least in the situation

of the oppressed man a kind of prefiguration by revolt of the socialist situation. (Ibid.: 171–2.)

Sartre insisted on what he called the 'dialectical', non-mechanical character of this conclusion, which he saw as partly implicit in the Trotskyist perspective. The victims of oppression are *persons*, not nails or billiard balls:

One oppresses a being with projects and oppression consists in the radical modification of all projects or alienation, and revolt as the projection of possible futures beyond the original alienation.

Moreover,

behind all the manifestations of oppression and the class struggle there appears in outline a future. The *whole* is present in every moment, as with Hegel. The classless society is present as a goal to be attained through every revolt, even in a sense through every oppression. It *defines* oppression. And finally, through all the concrete detail of struggles, of the tricks of the bourgeois class, and so on, oppression is always the *negative*, that is the *bad* social organisation which *prevents* the establishment of a rationally organised society. And the struggle against oppression is a struggle against the negative. That is, in marxist terms: negation of the negation. (Ibid.: 172.)

Stressing this view of the 'positive totality' as an ideal giving sense and meaning to the negative and destructive activity of revolution, Sartre concluded that 'the notions of value and, in a certain sense, of liberty are present in Trotskyism'. Its great difference with bourgeois morality lay elsewhere:

bourgeois morality like bourgeois law is *abstract*. It considers the person as pure subject of law and leaves aside concrete circumstances. For instance, the class struggle falls outside morality as pure empirical circumstance. Thereafter moral activity is presented as *independent* of historical circumstances. It is pure positivity and can always take place. Negativity properly so called can never colour moral action; it represents precisely immorality. Evil must not be defeated by Evil (negation of the negation) but by pure positivity. One combats oppression by charity, lying by telling the truth.

One must act to bring about the reign of morality not in relation to any concrete circumstance, but by appealing to abstractions (such as respect for persons). Such an attitude, Sartre argued, was a trap, hindering concrete action that might enable people to change their lot, action which is both negative and positive at once. This is why such morality focuses not on what one should do but on what, in all cases, one must not do, assuming that 'the Kingdom of Ends is eternal and that, by moral action, we participate in it' (ibid.: 173–4).

By contrast, for Trotsky, 'man's original aim is a concrete and historical aim: the Kingdom of Ends descends to earth, it is the socialist society to be realised'. Sartre stressed that what this involved was 'a concrete play of negations and affirmations', in particular 'the construction of a new society' and of 'a destructive instrument', in which the negative and positive aspects were linked:

One forges the destructive instrument by making it destroy. But precisely by giving the mass, so that it may destroy, that discipline, that cohesion, that self-denial, that self-confidence and that under-standing that makes of it the most formidable destructive instrument, one prepares it by this very fact for its positive role which is to become by itself the Kingdom of Ends; for the destructive instrument and the positive end are one and the same thing. Thus it is the means, at present, which makes concrete the end, which gives it, in some sense, body and individuality; or, if one prefers, it is within the means (the instrument) that one finds the end (preparation of the consciousness of the masses of the socialist society). (Ibid.: 174–5.)

In this way, 'the proletariat transforms itself into its own end', assimilating 'its cause to that of man'. This, already in 1947, was how Sartre saw marxism as a dialectic—'an effort to introduce morality into the concrete goal', the 'action of the whole on the parts and of the future on the present' (ibid.: 175). Central to this conception was an ambiguity, which Sartre saw as deriving from the Hegelian theory of liberty, between the 'whole' as ideal and as necessary future. In 1947, Sartre laid all the stress on the former and on an indeterminate future and the resulting importance of values. Trotsky, he thought, by rejecting certain means, accepted the 'existence of values and of liberty' (ibid.: 176).

The ambiguity of history was no less central to the thought of Maurice Merleau-Ponty, co-editor with Sartre of *Les Temps modernes*. They were critically minded but politically hopeful intellectuals preoccupied with the meaning of Soviet Communism, against the background of considerable Communist influence in post-Liberation France, above all in the cultural sphere. Needless to say, the orthodox Party intellectuals treated both of them with great suspicion and hostility. They both supported the doomed idea of a 'Third Force', unifying Europe under Socialist auspices; they were anti-Fascist, anti-capitalist, anti-American, anti-Gaullist, and anti-anti-Communist. (As Sartre later remarked: 'An anti-Communist is a dog, I do not depart from that, I shall never depart from it' [cited Caute 1973: 355].) They were not unaware of the nature of Stalinism in the late 1940s: as Sartre wrote in 1947:

Stalinism pollutes the working class, deprives it of all co-operative spirit and, by denying it democratic means within the Party, deprives it of confidence in itself and of educative experience. At the same time, Russian Stalinism appears as a threatening danger of war. (Sartre 1947–8: 170.)

In 1950 they jointly signed an article which acknowledged that Soviet citizens were deported without trial and that the number of detainees was between ten and fifteen million, and they denounced the police terror, the top-heavy bureaucracy, and the unsocialist salary differentials in the USSR; but anti-anti-Communism remained their dominant concern. As Sartre had written three years earlier, one should not attack Stalinism *'by all means'*: in particular, one should not, as Koestler believed, join one's 'forces with the reactionary forces (America, Gaullism, etc.) on the ground that only reaction can check the U.S.S.R. and the P.C.', for the 'triumph of those forces would not merely signify the liquidation of the P.C. but a reinforcement of the oppression of workers and a total loss of the confidence that that class has in itself' (Sartre 1947–8: 170–1). From the early fifties Sartre and Merleau-Ponty travelled in different directions. Sartre became a true fellow-traveller, in close proximity to the French Communist Party (though he never could climb aboard such a tight ship), until the Soviet intervention in Hungary, which he called a 'crime', denouncing the Party's 'repugnant lies' (see Caute 1973: 350) and thereafter taking up the causes of the peasant masses of the Third World, the *damnés de la terre*; while Merleau-Ponty moved towards liberalism and the politics of Pierre Mendès-France.

The debate over means and ends in which Sartre and Merleau-

Ponty took part, along with Camus and others, in post-war France, had as an essential reference-point Koestler's *Darkness at Noon*, which as *Le Zéro et l'infini* made publishing history after the Liberation, selling over 400,000 copies in France. It was, as Koestler later observed, 'the first ethical indictment of Stalinism published in post-war France; and as it talked the authentic language of the Party, and had a Bolshevik of the Old Guard for its hero, it could not easily be dismissed as "reactionary" and "bourgeois" ' (Koestler 1954: 403)—though of course it was. Indeed, it was held by at least one leading newspaper at the time to have been the single most important factor leading to the defeat of the Communists in the 1946 Referendum on the Constitution, which, had they won, would have given them nearly absolute control of the state (ibid.: 404).

Koestler's hero was, as he later wrote, 'a member of the Old Bolshevik guard, his manner of thinking modelled on Nikolai Bukharin's, his personality and physical appearance a synthesis of Leon Trotsky and Karl Radek' (ibid.: 394). The book was an exploration of the ultimate consequences of following the precept, 'the end justifies the means'. As Rubashov reflects at the end of the novel:

It was this sentence which had killed the great fraternity of the Revolution and made them all run amuck. What had he once written in his diary? 'We have thrown overboard all conventions, our sole guiding principle is that of consequent logics; we are sailing without ethical ballast'.

Perhaps the heart of the evil lay there. Perhaps it did not suit mankind to sail without ballast. And perhaps reason alone was a defective compass, which led one on such a winding, twisted course that the goal finally disappeared in the mist.

Perhaps now would come the time of great darkness. (Koestler 1940: 206.)

Koestler's exploration focused on the question: how were the hard-core Old Bolsheviks brought to confess to the absurd, and the essence of his answer comes in the passage where Rubashov is forced into submission by his second interrogator Gletkin—'a repellent creature, but he represented the new generation', a generation, Rubashov reflects,

which has started to think after the flood. It had no traditions, and no memories to bind it to the old, vanished world. It was a generation born without umbilical cord. . . . And yet it had right on its side. One must tear that umbilical cord, deny the last tie which bound one to the vain conceptions of honour and the hypocritical decency of the old world. Honour was to serve without vanity, without sparing oneself, and until the last consequence. (Ibid.: 149.)

Rubashov protests to Gletkin that, in opposing the Leader's policies, he did not act with counter-revolutionary intent, or as the agent of a foreign power, but in accordance with his conscience. Gletkin answers by quoting from his own writings: *'For us the question of subjective good faith is without interest. He who is in the wrong must pay; he who is in the right will be absolved. That was our law . . .'*; and again, *'It is necessary to hammer every sentence into the head of the masses by repetition and simplification. What is presented as right must shine like gold; what is presented as wrong must be as black as pitch . . .'* (ibid.: 187, 188). Finally, Gletkin rams his final argument home:

'Your faction, Citizen Rubashov, is beaten and destroyed. You wanted to split the Party, although you must have known that a split in the Party meant civil war. You know of the dissatisfaction amongst the peasantry, which has not yet learnt to understand the sense of the sacrifices imposed on it. In a war which may be only a few months away, such currents can lead to a catastrophe. Hence the imperious necessity for the Party to be united. It must be as if cast from one mould—filled with blind discipline and absolute trust. You and your friends, Citizen Rubashov, have made a rent in the Party. If your repentance is real, then you must help us to heal this rent. I have told you, it is the last service the Party will ask of you.

'Your task is simple. You have set it yourself: to gild the Right. To blacken the Wrong. The policy of the opposition is wrong. Your task is therefore to make the opposition contemptible; to make the masses understand that opposition is a crime and that the leaders of the opposition are criminals. That is the simple language which the masses understand. If you begin to talk of your complicated motives, you will only create confusion amongst them. Your task, Citizen Rubashov, is to avoid awakening sympathy and pity. Sympathy and pity for the opposition are a danger to the country.

Comrade Rubashov, I hope that you have understood the task which the Party has set you.'

It was the first time since their acquaintance that Gletkin called

Rubashov 'Comrade'. Rubashov raised his head quickly. He felt a hot wave rising in him, against which he was helpless. His chin shook slightly while he was putting on his pince-nez.

'I understand'. (Ibid.: 190.)

In 1947, Merleau-Ponty published his extraordinary *Humanism and Terror*, which argued, against Trotsky (whom it accused of 'rationalism'), that the future is essentially contingent and history ambiguous, delivering its verdict on the meaning of human actions only in the fullness of time. Preoccupied with the issues raised by Koestler and by Bukharin's trial, Merleau-Ponty's book, published just before the Prague coup and the Berlin crisis, exemplified what its author called a 'wait-and-see Marxism'—a position he was to abandon in the early fifties, when he stopped waiting and began to see. As Raymond Aron has remarked, 'the Moscow trials, as Koestler had presented them in *Darkness at Noon*, constituted the centre, the object and the occasion of [Merleau-Ponty's] essay. Rubashov tended to merge into Bukharin and the latter became an existentialist.' The trials became 'debates' (Merleau-Ponty's word) as though 'Vyshinksy and Bukharin discussed, as professors of philosophy, the respective roles of necessity and accident, of rationality and chance in the course of history. In the end, author and reader forget that the trials were fabricated in advance, their participants' roles assigned and all, judges and accused, recited scripts written in advance' (Aron 1983b: 312–13).

It was only too obvious, Merleau-Ponty observed, that Communism did not respect 'the rules of liberal thought'. The question was 'whether the violence it exercises is revolutionary and capable of creating human relations between men':

Does the violence in today's Communism have the same sense it had in Lenin's day? Is Communism still equal to its humanist intentions? (Merleau-Ponty 1947: xviii.)

His answer was that, while 'it would be difficult to maintain that [the Communist system] is moving towards the recognition of man by man, internationalism, or the withering away of the State and the realisation of proletarian power', none the less the Soviet Union should be given the benefit of the doubt;

given the assumption that human history was 'a totality moving towards a privileged state which gives the whole its meaning' (ibid.: 153), there was always a possibility that the Revolution would 'emerge from the Terror' (ibid.: 149), that indeed the Terror was a necessary condition for that emergence. In defence of this position, he cited the Trotsky of 1920 and accused the later, anti-Stalinist Trotsky of 'using the arguments of formal humanism against the dictatorship which now rules him when they once seemed false to him applied to the dictatorship he exercised' (ibid.: 85). It was, after all, 'difficult to delimit permissible Terror. There are all kinds of gradations between a Trotskyite dictatorship and a Stalinist dictatorship, and between Lenin's line and Stalin's line there is no difference that is an *absolute* difference' (ibid.: 91). Bukharin, Trotsky, and Stalin each 'imagined he was using [terror] to realise a genuinely human history which had not yet started but which provides the justification for revolutionary violence' (ibid.: 97), read into it a 'rational development', and drew from it 'a humane future'. Despite 'the ebb of revolution' (ibid.: xiii) and 'the problems of the USSR' (ibid.: xxiv), such a conclusion might be valid. Nevertheless, it was necessary to 'maintain the habit of discussion, criticism, research and the apparatus of social and political culture', and 'to preserve liberty while waiting for a fresh historical impulse which may allow us to engage it in a popular movement without ambiguity' (ibid.: xxiii). Marxism must aim at 'extrapolating, specifying and redirecting the spontaneous *praxis* of the proletariat along its proper path' (ibid.: 127).

It was a position, avoiding the choice between Communism and anti-Communism, that Merleau-Ponty was to retract (oddly enough in reaction to the Korean war), arguing in 1955 that, although after World War II 'wait-and-see Marxism' had 'objective conditions', it was 'nothing more than a dream, and a dubious dream' (Merleau-Ponty 1955: 230). It was the dream of the classical fellow-travelling intellectual, offering fellowship (but not allegiance) on the assumption that, whatever the appearances, Official Communism was (or might be) travelling, albeit by 'dialectical detours', towards true Communism. Indeed, Merleau-Ponty expressed this position most succinctly in 1948 against Claude Lefort's view that the USSR stands *accused*. For us, with its grandeurs and its horrors, it is an enterprise that has had a breakdown. It is necessary to say if one is Communist or not, and we say we are not. But the prosecutor's tone appears to us to be misplaced in a world that is nowhere innocent and does not appear to be governed by an immanent rationality. (*Les Temps modernes*, 29, 1948: 1516.)

It was a dream quite unrelated to the real nightmare of Soviet life in the thirties and forties with its mass killings and deportations, purges and show trials, its climate of fear, corruption, and mutual suspicion, all engendered and manipulated by a paranoid and cynical leadership in the process of extending its sway and methods across Eastern Europe. But it was a dream that exerted great influence on the European left, especially in France, among whom it was not Merleau-Ponty's book but Raymond Aron's criticism of it that caused a scandal at the time.

Humanism and Terror was, in part, a striking restatement of some of the classical positions towards moral questions that we have seen to be typical of Marx and of marxists. It was, and remains, striking partly because it restated them in so explicit and systematic a fashion, and partly because it did this so late and in (ambiguous) defence of a system of terror whose relation to humanism could only be discerned by an intellectual unaware of, or blind to, its nature and scope (who could write that 'in the Soviet Union, violence and deception are official while humanity is in daily life; in the democracies on the other hand principles are humane, while deception and violence are found in practice' [cited Cooper 1979: 83]).

It made the standard criticisms of liberal (especially Kantian) ethics as being and relying upon 'abstract distinctions of liberal thought' (Merleau-Ponty 1947: 32), and as a form of ideological deception. The 'liberal ideal of justice plays a role in the operation of conservative society' (ibid.):

To tell the truth and to act out of conscience are nothing but alibis of a false morality; true morality is not concerned with what we think or what we want but obliges us to take an historical view of ourselves. (Ibid.: 103–4.)

Either

one wants to do something, but it is on condition of using violence— or else one respects formal liberty and renounces violence, but one can only do this by renouncing socialism and the classless society, in other words by consolidating the rule of 'Quaker hypocrisy'. (Ibid.: 107.)

Koestler was, according to Merleau-Ponty, engaged in rehabilitating 'moralism and the ''beautiful soul'' ', posing 'the problem in pre-Marxist terms':

To confront Rubashov with the Christian absolute 'Yea' or 'Nay', or Kant's moral imperative, simply shows that one refuses to face the problem and falls back upon the attitudes of the holy will and the pharisee. It is necessary from the start to recognise *as a moral claim* the Communist's preoccupation with the role of objective factors and his wish to look upon himself from a standpoint both within and outside of history. (Ibid.: 22.)

And such a standpoint, Merleau-Ponty made clear, was anti-utopian:

Marxism very consciously distinguishes itself from utopianism by defining revolutionary action not as the adoption of a certain number of ends through reasoning and will, but as the simple extrapolation of a *praxis* already at work in history, of a reality that is already committed, namely, the proletariat. . . . [It makes] the creative decision to pass beyond this chaos, through the universal class that will relay the foundations of human history. (Ibid.: 126.)

Humanism and Terror went beyond restating and defending this standpoint: it was also, in part, an application of it, to the critique of Koestler and to the interpretation of Bukharin's trial. Koestler was 'far from giving a true presentation' of the Moscow Trials: such a presentation would show 'not the Yogi at grips with the Commissar—moral conscience at grips with political ruthlessness, the oceanic feeling at grips with action, the heart at grips with logic, the man without roots at grips with tradition . . .' (ibid.: 62). On the contrary,

the true nature of tragedy appears once *the same man* has understood that he can disavow the objective pattern of his actions, that he is what he is for others in the context of history, and yet that the motive of his actions constitutes a man's worth as he himself experiences it. . . . We are no longer dealing with Rubashov who gives in unconditionally once he falls back into the comradeship of the Party and denies everything including his own past when he hears the cries of Bogrov; we have instead Rubashov who adopts history's viewpoint on himself, and who works for his own condemnation from the side of history while defending his revolutionary honour. (Ibid.: 62–3.)

Merleau-Ponty's reading of Bukharin's trial was that Bukharin, tragically, was 'a man at grips with external forces *with which he is secretly allied*'—that 'between Bukharin and the court, while there was no express agreement, there was at least the tacit bond that both were marxists' (ibid.: 48). Merleau-Ponty did not seem to consider Bukharin to be guilty *as charged* (and for this the Communist critics of his book could not forgive him), but this was not the point: Bukharin himself showed that

in the situation given and in the logic of the struggle, his evaluation was in fact counter-revolutionary, and he is therefore guilty of historical treason. It is quite evident that Bukharin is no Fascist. He took precautions against the Bonapartist tendencies that he suspected in military circles. What is true is that in the battle over collectivisation, the opposition could only rely on the Kulaks, Mensheviks and the remaining elements of the Social Revolutionaries—could only overthrow the Party leadership with their help—that it would have to share power with them and that thus in the end there were 'elements of Caesarism' involved. No, it is quite obvious that Bukharin was not linked with White Guard emigré Cossack circles. But plainly he was interested in the Kulak opposition. He kept informed on the Kulak revolts through friends coming from the Northern Caucasus or from Siberia, who in turn got their information from Cossack circles. Consequently, he accepts responsibility for these revolts. . . . The Prosecutor's role is to reveal Bukharin's activity on the plane of history and objectivity. Bukharin regards this interpretation as legitimate; he only wants it to be known that it is an interpretation and that it is only from a certain standpoint that he was linked with the Cossacks. (Ibid.: 52–3.)

From this standpoint, the Moscow Trials appeared as 'not an act of timeless justice but a phase in the political struggle and an expression of the violence in history' (ibid.: 30–1); they were 'revolutionary trials presented as if they were ordinary trials' and 'only make sense between revolutionaries, that is to say between men who are convinced that they are *making history* and who consequently already see the present as past and see those who hesitate as traitors' (ibid.: 29). Thus:

Bourgeois justice adopts the past as its precedent; revolutionary justice adopts the future. It judges in the name of the Truth that the Revolution is about to make true; its proceedings are part of a *praxis* which may well be motivated but transcends any particular motive.

The trials were not concerned with 'whether the accused's

motives and intentions were honest or dishonest'; only with 'whether in effect his conduct, understood from the standpoint of the collective *praxis*, is revolutionary or not' (ibid.: 28). For,

political acts are to be judged not only according to their meaning for the moral agent but also for the sense they acquire in the historical context and dialectical phase in which such acts originate. Moreover it is impossible to see how a Communist could disavow this approach, as it is essential to Marxist thought. In a world of struggle—and for Marxists history is the history of class struggles—there is no margin of indifferent action which classical thought accords to individuals; for every action unfolds and we are responsible for its consequences. (Ibid.. 33.)

From this Merleau-Ponty drew the remarkable conclusion that 'in a period of revolutionary tension or external threat there is no clear-cut boundary between political divergences and objective treason' (ibid.: 34). Violence will not be driven out of history by 'locking ourselves within the judicial dream of liberalism' (ibid.: 35); historical responsibility 'transcends the categories of liberal thought' (ibid.: 43). In short, for Merleau-Ponty, because the opposition 'weakened the USSR', the Moscow Trials were to be seen (like those of Nazi collaborators) as 'the drama of subjective honesty and objective treason' (ibid.: 44).

What Merleau-Ponty proposed, in effect, was a kind of ultra-consequentialism, in which the very meaning of an action is determined by its long-term results. According to this view, it is from 'the perspective of the future' (ibid.)—in this case 'the Stalinist perspective on Soviet development' (ibid.: 28)—that an objective judgement can be made on a social movement and on a man's character and actions. Such a judgement is, of course, subjectively made, but what settles its objectivity is 'the role he plays in a network of events': 'Thus is it possible to have to answer for acts of treason without having intended them' (ibid.: 58). What counted in assessing Bukharin's actions and his responsibility for them was 'the objective pattern of his actions' and where they led 'in the end'. Not only is the meaning of an act for the moral agent of no account in assessing its justification or the agent's responsibility; these questions are to be decided in the long perspective of a projected, or imagined, future, which is seen as having 'absolute validity' and as giving revolutionaries 'the assurance of understanding of what they are living through' (ibid.: 28–9). In this astonishing work, Merleau-Ponty was prepared to gamble on the 'probability' (ibid.: 28) that that assurance was justified, and that the perspective he brought to bear on the Moscow Trials—and more generally the Stalinist terror—provided 'a new foundation for historical truth' in 'the proletariat's self-recognition and the real development of the revolution' (ibid.: 18).

Koestler's comment on *Humanism and Terror* was that 'Professor Merleau-Ponty, the successor to Henry Bergson's chair at the College de France, published a remarkable book to prove that Gletkin was right.' The book, he added,

defends every measure of the Soviet regime, including the Stalin-Hitler pact, as Historical Necessity, condemns Anglo-American policy as Imperialist Aggression, and regards criticism of the Soviet Union as an implicit act of war. It is an almost classic example of the controlled schizophrenia of the closed system, provided by the foremost academic exponent of the French Marxist-Existentialist school. (Koestler 1954: 404.)

It also classically exemplifies the point to which a theoretically (if not factually) well-schooled thinker could go when reasoning within the structure of thought that it has been this book's aim to exhibit and analyse.

Conclusion

It is strange, by the way, that Marxists, who pride themselves on their realism, should habitually regard the Present as merely the mean entrance-hall to the spacious palace of the Future. For the entrance-hall seems to stretch out interminably; it may or may not lead to a palace; meanwhile it is all the palace we have and we must live in it.

(Macdonald 1946: 98.)

The calm confidence—or ecstasy—of the political leader who sends masses of humanity to their death for the sake of a shining distant future is indeed abominable. Equally abominable is the complacency of those liberals willing to rain terror from the skies while they prate about the virtues of pragmatic gradualism. Repulsive moral certainty is not limited to fanatics, while to refuse to act in clear circumstances where the consequences are apparent can have its own tragic results.

(Moore 1972: 27–8.)

This book has been about marxism as a theory, a way of interpreting the world with a view to changing it, and not primarily about what has been done in its name or what it has been invoked to justify. Since 1917, its development has been indissolubly linked to the fates of the various social experiments conducted in its name and under conditions it never envisaged—extreme scarcity, backwardness, war, and isolation. They have all occurred (with the quasi-exception of the Soviet absorption of Eastern Europe after Yalta) in backward conditions far from the centres of advanced capitalism. Since World War II nearly twenty countries have acquired regimes professing adherence to marxism, all in the Third World. Their marxism, like the original Russian and Chinese exemplars, is syncretic, an amalgam of bibliocentric orthodoxy and indigenous traditions, concocted under conditions highly adverse to both socialism and democracy.

Revolutionary leaders, struggling, under such adverse conditions, to maintain social order, build a state, and achieve economic development, while threatened with hostilities and dangers from within and without, will select and interpret whatever ideological justifications lie at hand to legitimate their actions and policies. In the post-revolutionary period, these are in turn transformed into a state ideology that henceforth both limits and adapts to the policies of the ruling stratum. Thus in the USSR, 'triumphant Stalinism twisted and up-ended virtually every Marxist ideal and rudely contradicted Lenin's vision in 1917 of destroying bureacracies and standing armies', while in China, only after the crushing of the predicted proletarian risings in the cities, and the rise of peasant-oriented movements in the military base areas in the countryside, 'did "Maoist" doctrine develop to sanctify and codify what had been done. Thereafter epicycles were always added to the base model wherever necessary to justify practical detours on the road to national power' (Skocpol 1979: 171). Marxism as a state ideology has also absorbed accretions from pre-revolutionary traditions and of course from the liberal and capitalist West, not least, and in considerable measure, the language of moralism and human rights, itself largely deriving from an earlier revolutionary tradition of which marxism is a proud and self-conscious heir.[1] There is no finer exemplar of liberal constitutionalism than Stalin's Constitution of 1936.

Many marxists, ever since Trotsky, have sought to account for the failures of these experiments in such a way as to save the theory, indeed often *in terms of* the theory. Of course, there has been no shortage of those who sought to deny that the experiments *were* failures: as observers and often visitors, they ignored or underplayed or whitewashed, or else defended or even celebrated their systematic and large-scale deception, violence, irrationality, and cruelty (Caute 1973). Equally, of course, there has been no shortage of those who have been content to see such failures as enough to discredit the theory, and indeed the very idea of socialism itself.

It is no part of this book's aim to lend any support to this last

---

[1] The French revolutionary tradition was itself ambiguous between (at least) the inheritance of Montesquieu, for whom the *esprit général* embodied competing sectional interests that were to be balanced within a constitutional pluralism, and that of Rousseau, for whom the *volonté général* would transform such 'factional' interests into a virtuous consensus within a rejuvenated ideal republic. Robespierre and Saint Just were plainly Rousseauists (see Hampson 1983).

position. On the contrary, it is intended as a contribution to socialist free-thinking—freed in particular from some of the illusions and blindspots of what has been the overwhelmingly dominant socialist tradition in this century. Its argument, to state it shortly, is this: that marxism has from its beginning exhibited a certain approach to moral questions that has disabled it from offering moral resistance to measures taken in its name; in particular, that, despite its rich view of freedom and compelling vision of human liberaton, it has been unable to offer an adequate account of justice, rights, and the means-ends problem, and thus an adequate response to injustice, violations of rights, and the resort to impermissible means, in the world we must live in. Let me try to draw together the strands of this argument, by, first, characterizing marxism's approach to morality; and, second, by identifying its major weaknesses and strengths.

The marxian *oeuvre*, published and unpublished, was, to say the least, an ambiguous bequest to the marxist tradition; and nothing could show this more clearly than the various positions reviewed in the previous chapter. For the marxist and *marxisant* thinkers there considered, locked into such bitter dispute with one another's responses to pressing contemporary political problems, saw themselves as interpreting and responding to them according to the marxist method. And this self-understanding was, in each case, plausible: Rosa Luxemburg's belief in the effectiveness of self-transforming, self-educating, and self-correcting revolutionary mass action; Kautsky's belief in a majoritarian, constitutional, and minimally coercive evolution to socialism in advanced European conditions; Trotsky's ruthlessly instrumental approach to the class war, avoiding the moralistic and legalistic 'class deceptions' of the enemy; Lukács's belief that the proletariat's goal is written into history and that its standpoint yields 'correct' answers to tactical and ethical questions; Sartre's notion of the future classless society as a 'positive totality' giving sense and meaning to oppression and the struggle against it, and of the revolutionary class as transforming itself into its own end; and Merleau-Ponty's view of history as ambiguous but correctly interpretable, from the proletarian standpoint, as showing the need for 'dialectical detours' to achieve a future of 'humanity'—all apply—without distorting the texts—aspects of marxian doctrine. The fact that Luxemburg and Kautsky, and later Serge, criticized, while Lukács defended, Lenin's and Trotsky's practice; that Lenin and Trotsky came to denounce Kautsky for betrayal and moralism; that Trotsky was, with his old ally and latter-day critic Serge, later the most consistent and bitter critic of Stalinism, with which Lukács came to ambiguous terms, and which both Sartre and Merleau-Ponty for a time defended, while Sartre praised and Merleau-Ponty criticized Trotsky as a moralist—all this shows most vividly that the marxist method, even in these most practised hands, did not yield convergent answers to practical problems.

Their answers do, however, for the most part, reveal certain common crucial presuppositions and implications. They support and exemplify the central claim of this book: that marxian and marxist thought about the means and ends of action, and more generally about morality, has a certain distinctive structure. It is, in brief, a form of consequentialism that is long-range and perfectionist.

By 'consequentialism' I mean a theory which judges actions by their consequences only, and requires agents to produce the best available outcome, all things considered: it relates the right and good by holding that it is always right that agents should act so as to bring about the best outcome overall (see Scheffler 1982). Plainly, consequentialism contrasts with deontological theories, which standardly hold that it is sometimes *wrong* to produce the best outcome overall, and *right* not to do so, by imposing 'side-constraints' or 'agent-centred restrictions'. According to such theories, the 'right' is not related instrumentally to the 'good' but is otherwise justified, e.g. by reference to the 'moral law' or to some notion of personal integrity or respect for persons, or to the will of God. It therefore comes as no surprise that marxism is deeply and unremittingly anti-deontological: hence the systematic hostilities we have traced among the orthodox to Kant and kantianism. Or, to put it another way, relaxation of such hostilities represents a deep form of revisionism. For marxism and orthodox marxist thought require an exclusive, single-minded preoccupation with the attainment of emancipation. 'He who fights for Communism', wrote Brecht, 'has of all virtues only one: that he fights for

Communism' (Brecht 1929–30: 13). The purportedly 'eternal', 'universal', and 'abstract' principles adduced by deontological theories are, from its viewpoint, both without foundation, and, if applied, irrelevant to and obstructive of the consequences it requires action to promote.

The point is well brought out in Brecht's play, *Die Massnahme* (The Measures Taken). For Brecht, as Martin Esslin has written, 'the relation between ends and means always remained a vital problem'. This play, written in 1929–30, 'is an exact and horrifying anticipation of the great confession trials of the Stalinist era. Many years before Bukharin consented to his own execution in front of his judges, Brecht had given that act of self-sacrifice for the sake of the party its great, tragic expression' (Esslin 1959: 154, 144). In the play, 'Four Agitators' decided to shoot a soft-hearted 'Young Comrade' who has flouted Party discipline, to relieve suffering because 'misery cannot wait', thus putting them, and the revolution, in danger:

And so we decided: we now
Had to cut off a member of our own body.
It is a terrible thing to kill.
We would not only kill others, but ourselves as well, if the need arose.
For violence is the only means whereby this deadly
World may be changed, as
Every living being knows.
And yet, we said
We are not permitted to kill. At one with the
Inflexible will to change the world, we formulated
The measures taken.

To which the 'Control Chorus' responds:

It was not easy to do what was right.
It was not you who sentenced him, but
Reality.

For 'what is needed to change the world' are

Anger and tenacity, knowledge and indignation
Swift action, utmost deliberation
Cold endurance, unending perserverance
Comprehension of the individual and comprehension of the whole:
Taught only by reality can
Reality be changed. (Brecht 1929–30: 32–3, 34.)

But how does marxism conceive of the consequences by which, through its understanding of 'reality' and 'comprehension of the whole', it judges actions and policies? What sort of consequentialism does it, in turn, advance? Though Marx and later marxists (notably Lenin) sometimes speak in utilitarian accents, there is little room to doubt that, as Chapter Five has shown, the essential message is perfectionist: the best over-all outcome is the maximal realization of human powers, of many-sided individuality, in community, the attainment of a society in which the full and free development of every individual forms the ruling principle. There is no discussion, either in Marx or in the marxist tradition, of whether, how, and why such an ideal might diverge from the utilitarian, such a question appearing self-evidently utopian. Thus, although Marx and later marxists were not perfectionists *rather than* utilitarians, they are best interpreted as the former, not the latter. It is, moreover, an ideal that is decidedly long-range in nature: however immanent and imminent its realization might have seemed in revolutionary times, such as the 1840s or after the October Revolution, commitment to its promotion is clearly (as utilitarianism tends not to be) a commitment to the long haul.

It is a commitment that has been absolutely central to marxism in theory and practice. As we have seen, Marx wrote that the working-class know that

in order to work out their own emancipation, and along with it that higher form to which present society is irresistibly tending by its own economical agencies, they will have to pass through long struggles, through a series of historical processes, transforming circumstances and men. They have no ideals to realise, but to set free elements of the new society with which old collapsing bourgeois society is pregnant. (Marx 1871a: 523.)

Such a commitment is far from the Christian precept: 'Take therefore no thought for the morrow: for the morrow shall take thought for the things of itself. Sufficient unto the day is the evil thereof.' To reject its exclusive focus on the goods of the morrow is to reject an idea essential to the marxist canon. It is (to change the image to Dwight Macdonald's cited in this chapter's epigraph) to abandon the quest for the spacious palace at the end of the entrance-hall in which we must live, on the grounds that 'we shall live in it better and even find the way to the palace (if there is a palace), if we try living in the present instead of in the future' (Macdonald 1946: 98).

To say this is the most decisive form of Revisionism. And this is exactly what Eduard Bernstein did say when he famously remarked that 'To me that which is generally called the ultimate aim of socialism is nothing, but the movement is everything.' In the last chapter of his *Evolutionary Socialism* (entitled 'Ultimate Aim and Tendency—Kant against Cant'), Bernstein interpreted the passage I quoted above from Marx in his own way, rejecting the first part of its last sentence as 'self-deception'. For Bernstein,

Whether [the working class] sets out for itself an ideal ultimate aim is of secondary importance if it pursues with energy its proximate aims. The important point is that these aims are inspired by a definite principle which expresses a higher degree of economy and of social life, that they are an embodiment of a social conception which means in the evolution of civilisation a higher view of morals and of legal rights. (Bernstein 1899: 222.)

In short, such a rejection of an exclusive morality of emancipation implies a return of the morality of *Recht*.

How, finally, are we to assess marxism's morality of emancipation, as an action-guiding theory, a way of seeing and of not seeing? I have in this book focused on two kinds of defect: its generation of illusions and its systematic blindness.

The illusions come into view when we ask the question: if actions (or at least the actions of revolutionaries) are to be judged for their rightness by their contribution to long-range perfectionist consequences, how is *that* judgement to be made? Here Dewey's criticism of Trotsky touches a raw nerve. For certain key features of marxist thought have all too often led marxists away from a comparative assessment of the probable long-term outcomes of alternative means. First, there is the anti-utopian argument that the envisaging of long-term goals is illegitimate, diversionary, and in any case impossible: the working class has 'no ideals to realise', merely to 'set free elements of the new society with which the old collapsing bourgeois society is pregnant'. Second, there is the wholly unwarranted assurance (so often revealed by that telling word 'correct') with which marxists have approached the prediction of future consequences; as the Soviet joke has it, 'The future is certain; it is only the past that is unpredictable.' The scientism of the Second and Third Internationals (not altogether absent from Marx and Engels) provided one such false warrant. Another, more sophisticated, is the presupposition well expressed by Lukács, that the proletarian standpoint (interpreted, *he* assumed, by the Party) provides a kind of privileged access to both past and future: that it yields *knowledge* of the dynamics and direction of history and of the general (though not the specific) direction of the future. In short, a class-based perspective, and the necessity of class struggle, was assumed to be inseparable from knowledge itself, and thus from objectively based judgement. In these ways, marxists have led themselves to believe that what Barrington Moore calls 'clear circumstances where the consequences are apparent' are far more frequent than they are. Merleau-Ponty was, of course, quite free of this illusion, though he fell for the third, which runs through the thought of most of the marxists we have considered and is once more accurately and well expressed by Lukács, namely the presupposition that the 'ultimate objective' of human emancipation is immanent in world history, and that the proletariat's 'mission' is to bring it to fruition.

In short, these three illusions display marxist consequentialism as an irrational endeavour. Because it inhibits the specification of its ultimate aim, while presuming to foresee the future, in which its eventual realization is somehow guaranteed, it forswears both the clarification of the long-term consequences by which alternative courses of action are to be judged and, as Dewey put it, the 'open and unprejudiced' examination of those alternatives.

In what respects is marxism morally blind? What does the marxist version of consequentialism rule out or ignore in the assessment of human action and character? A general answer is: all that it holds to be irrelevant to the project of human emancipation; and, in particular, the interests of persons in the here and now, both victims (intended and unintended) and agents, in so far as these have no bearing on that project.

This is perhaps why marxism has had so little to say about the sphere of personal morality and interpersonal relationships, which is generally, and popularly, conceived as the heart, even the essence of morality: it identifies the obstacles to emancipation at the macro level; they are social and historical, rather than psychological and anthropological, to be found within social structure and historical process rather than in the processes of human interaction or the nature of man. It is also why it has been at best ambivalent about the domain of *Recht*, tending to see present injustices and violations of rights, not *as* injustices and violations of rights, but rather as either obstacles or means to future emancipation. And it is why marxism has never come properly to grips with the means-ends issue, and the problem of dirty hands. For it is resistant to the perception that moral conflict is at issue here: that in pursuing the course of action with the best overall consequences, we may do what is wrong. Armed with an illusory certainty about the direction of history, while refusing to reflect upon its supposed destination, it has no basis for resisting any measures taken that appear to promote such consequences.

A comparison with another form of consequentialism, utilitarianism, may be helpful here. Utilitarians too, ever since Bentham spoke of rights as 'nonsense' and of natural and imprescriptible rights as 'nonsense on stilts', have had difficulties with the concept of rights. They have also had difficulties with the notion of justice, their critics holding that the consistent utilitarian must allow injustice where the promotion of utility requires it. And on the issue of dirty hands, they will also tend to deny that any dilemma exists, since if the appropriate calculations show what is the right thing to do in utilitarian terms, then that is the right thing to do. Of course, they may well then go on to give a utilitarian explanation and justification for the moral rules, for the principles of justice and the rights and obligations by which we ordinarily live, and further for the sense of obligation we attach to these and for the guilt we feel when we violate them. But, in any given case, there is a correct answer to the question 'what is it right to do?', which might require us to override the constraints of ordinary morality, and in that case to do so would not be wrong.

Marxist consequentialism is like utilitarianism in offering an explanation for the constraints of ordinary morality (though it offers a different explanation), but unlike it in offering little scope for justifying them: on the contrary, it holds that such constraints are likely to be class deceptions, lying in ambush to trap the unwary. Moreover, the justification for such constraints that a utilitarian can offer will incorporate in its calculations benefits to agents in the here and now and the foreseeable future. For the marxist, such benefits are, in themselves, irrelevant. The long-term character of marxist consequentialism, focusing on future benefits of future persons, make it markedly less sensitive than even utilitarianism to the moral requirement of respecting the interests of persons in the present and immediate future.

In particular the mainline marxist tradition has no room for the crucial thought that, even though overall good may be served by the violation of rights and the committing of injustice, an uncancelled moral wrong or harm is done to the victims (Williams 1978): as Brecht's Control Chorus put it, 'It was not easy to do what was right.' It cannot accommodate Macchiavelli's thought that sometimes, as with Romulus and Numa, 'when the act accuses, the result excuses' (since for the act to accuse, moral standards must be deemed to apply: the accusation stands); or Max Weber's view of the political leader as riven by an irreconcilable conflict between the demon of politics and the god of love; or Camus's and Koestler's idea that those who go beyond certain limits, even with consequential justification, are criminals and must pay for their crimes. Nor has it a place for the broadly kantian thought that the interests of all persons are equally owed a consideration, just *because* they are persons, that requires just that sort of abstraction that marxism strenuously resists—an abstraction from the roles people play (capitalist, worker, etc.) and the categories they exemplify (class or party enemy), in order to see their world from within, from their separate and individual points of view. This is the thought that Serge expressed, as taking precedence before all tactical considerations:

1. *Defence of man. Respect for man.* Man must be given his rights, his security, his value. Without these there is no Socialism. Without these, all is false, bankrupt and spoiled. I mean, man whoever he is, be he the meanest of men—'class enemy', son or grandson of a bourgeois, I do not care. It must never be forgotten that a human being is a human being. (Cited Serge 1942–3: 282–3.)

Nor, finally, does it have room for the further thought that only if the politics and practice of a movement or system inculcates and encourages sensitivity to these thoughts, will it have any chance at all of bringing into being a socialism that is worth fighting for and defending.

What, finally, can be said of the strengths of marxism's view of morality? It is illusory and partially blind, but it is also extraordinarily penetrating, in at least two ways.

First, as we have seen in Chapter Five, it offers a conception, a way of interpreting the concept, of freedom, and of the constraints upon or obstacles to it, that is far deeper and richer than negative and classical liberal views. As an account of the sources of unfreedom and as a vision of emancipation (albeit ambiguous in the ways we have indicated and ungrounded in a determinate view of the human subject), it cannot be ignored by those who profess to take liberty seriously, which must include at least those who take socialism seriously.

Second, and most challengingly, it raises, as we have repeatedly seen, some deep and unanswered questions about the morality of *Recht*, which non-marxists have yet to answer. How can the abstractions of *Recht* be grounded? What *are* the generically human interests that underpin talk of human rights? Why should they have more than merely local standing? How and on what basis can we abstract 'persons' from their positions and social roles, adopting 'their' point of view, so as to view them from the standpoint of justice? If we attempt to engage in such abstractions, do we not end up with an attempt to isolate what cannot be isolated? What gives moral constraints any more than merely local applicability and force? Are they not, as Trotsky and Sartre suggest, merely traps to hinder the wretched from improving their lot? Why should the question 'What is not to be done?' have *any* general and rationally compelling answer, imposing obligations upon us? Most important of all, can a theory of justice and of rights be developed which incorporates the insight and vision of marxism's conception of freedom? Though marxism sees no need to answer any of these questions, it inescapably raises them, both by its critique of morality in theory and by its moral record in practice.

Bibliography

Works are dated either by first publication or, when posthumous, by their date of composition. Page references in the text of this book are to the edition or, when a quotation is translated, to the translation indicated below. When a translation has been amended, this is indicated by the initials 'S.L.'

Adler, M. (1925), *Kant und der Marxismus*. E. Laub'sche Verlagsbuchhandlung, Berlin.

Adorno, T. (1951), *Minima Moralia*. Suhrkamp Verlag. Trans. E. F. N. Jephcott. New Left Books, London, 1974.

Allen, D. (1973), 'The Utilitarianism of Marx and Engels', *American Philosophical Quarterly*, 10.

Anderson, P. (1980), *Arguments within English Marxism*. Verso, London.

Aron, R. (1983a), 'The Life and Death of Arthur Koestler: A Writer's Greatness', *Encounter*, no. 357 (July–Aug.).

——(1983b), *Mémoires*. Julliard, Paris.

Avineri, S. (1972), *Hegel's Theory of the Modern State*. Cambridge University Press, Cambridge.

Bauer, O. (1906), 'Marxismus und Ethik', *Neue Zeit*, XXIV. 2. (Critique of Kautsky, 1906a.) Translated in part in Bottomore and Goode 1978.

Berlin, I. (1958), 'Two Concepts of Liberty', reprinted in *Four Essays on Liberty*. Oxford University Press, Oxford, 1969.

Bernstein, E. (1899), *Die Voraussetzungen des Sozialismus und die Aufgaben der Sozialdemokratie*. Dietz, Stuttgart. Translated by Edith C. Harvey (1909) as *Evolutionary Socialism: A Criticism and Affirmation*, reprinted by Schocken Books, New York, 1961.

Bloch, E. (1918), *Geist der Utopie*. Original edition Volume 16 and Revised edition (1923) Volume 3 of Bloch's *Collected Works*, Suhrkamp, Frankfurt am Main.

——(1954–9), *Das Prinzip Hoffnung*. Volume 5 (3 parts) of Bloch's *Collected Works*, Suhrkamp, Frankfurt am Main.

Bottomore, T. and Goode, P. (1978), *Austro-Marxism*. Clarendon Press, Oxford.

Brecht, B. (1929–30) *Die Massnahme*. Translated as *The Measures Taken* in *The Measures Taken and Other Lehrstücke*. Eyre Methuen, London, 1977.

——(1938), 'An die Nachgeborenen'. Translated as 'To Those Born Later' in *Bertolt Brecht Poems 1913–1956*, edition by John Willett, with Ralph Manheim with the co-operation of Erich Fried. Methuen, London, 1976.

Brenkert, G. G. (1975), 'Marx and Utilitarianism', *Canadian Journal of Philosophy*, 5.

——(1983), *Marx's Ethics of Freedom*. Routledge and Kegan Paul, London.

Buchanan, A. E. (1982), *Marx and Justice: The Radical Critique of Liberalism*. Methuen, London.

Campbell, T. (1983), *The Left and Rights: A Conceptual Analysis of the Idea of Socialist Rights*. Routledge and Kegan Paul, London.

Caute, D. (1964), *Communism and French Intellectuals 1914–1960*. Andre Deutsch, London.

——(1973), *The Fellow Travellers*. Weidenfeld and Nicolson, London.

Chamberlin, W. H. (1935), *The Russian Revolution, 1917–1921*. 2 Vols. Grosset and Dunlap, New York, 1965.

Cliff, T. (1975), *Lenin*, Volume 1: *Building the Party*. Pluto, London.

Cohen, G. A. (1981), 'Freedom, Justice and Capitalism', *New Left Review*, no. 126 (Mar.–Apr.).

——(1983), Review of Wood 1981, *Mind*, vol. XCII, no. 367 (July).

Cohen, M., Nagel, T., and Scanlon, T. (1980), *Marx, Justice and History*. Princeton University Press, Princeton, New Jersey.

Cohen, S. F. (1974), *Bukharin and the Bolshevik Revolution: A Political Biography 1888–1938*. Wildwood House, London.

Colletti, L. (1975), Introduction to Karl Marx, *Early Writings*. Penguin Books in association with *New Left Review*.

Cooper, B. (1979), *Merleau-Ponty and Marxism: From Terror to Reform*. University of Toronto Press, Toronto.

Cornell, D. (1984), 'Should a Marxist believe in Human Rights?', *Praxis International*, 4.1. (Comment on Lukes 1982.)

Davis, J. C. (1981), *Utopia and the Ideal Society. A Study of English Utopian Writing 1516–1700*. Cambridge University Press, Cambridge.

Dewey, J. (1938), 'Means and Ends', *The New International*, Aug. 1938. Reproduced in *Their Morals and Ours. Marxist versus Liberal Views on Morality*. Four essays by Leon Trotsky, John Dewey, and George Novack. 4th edn., Pathfinder Press, New York, 1969.

Dewey, J. *et al.* (1969), *The Case of Leon Trotsky*. Merit, New York.

Dworkin, R. (1977), *Taking Rights Seriously*. Duckworth, London.

Elster, J. (1985), *Making Sense of Marx*. Cambridge University Press, Cambridge.

Engels, F. (The *Collected Works* of Marx and Engels, published in English translation by Lawrence and Wishart, London, from 1975 are indicated by MECW; the *Selected Works* in two volumes published by the Foreign Languages Publishing House, Moscow, in 1962 by SW; and the *Selected Correspondence* published by the Foreign Languages Publishing House, Moscow (no date) by SC.)

Engels, F. (1845a), Speeches at Elberfeld, MECW 4.

—— (1845b), *The Condition of the Working Class in England*, *MECW 4*.

—— (1872–3), *The Housing Question*, SW 1.

—— (1874), 'On Authority', SW 1.

—— (1875), Prefatory Note to *The Peasant War in Germany*, SW 1.

—— (1877–8), *Anti-Dühring*, Foreign Languages Publishing House, Moscow, 1959.

—— (1880), *Socialism: Utopian and Scientific.* Three chapters from *Anti-Dühring* revised to form a short book, SW 2.

—— (1883), Speech at the Graveside of Karl Marx, SW 2.

—— (1884), Preface to the first German edition of Marx, *The Poverty of Philosophy*. Progress Publishers, Moscow, rev. edn., 1975.

—— (1885), *On the History of the Communist League*, SW 2.

—— (1890), Letter to C. Schmitt, 5 August 1890, SC.

Esslin, M. (1959), *Brecht: A Choice of Evils.* 3rd rev. edn., Eyre Methuen, London, 1980.

Feinberg, J. (1973), *Social Philosophy*. Prentice-Hall, Englewood Cliffs, N. J.

Finnis, J. (1980), *Natural Law and Natural Rights*. Clarendon Press, Oxford.

Fromm, E. (ed.) (1967), *Socialist Humanism*. Allen Lane: The Penguin Press, London.

Garton Ash, T. (1983), 'The Poet and the Butcher', *Times Literary Supplement*, 4,210 (9 Dec.).

Geras, N. (1976), *The Legacy of Rosa Luxemburg*. New Left Books, London.

—— (1983), *Marx and Human Nature: Refutation of a Legend*. NLB and Verso, London.

Goldman, Emma (1925), *My Disillusionment in Russia*. C. W. Daniel, London.

Green, M. (1983), 'Marx, Utility and Right', *Political Theory*, 11. (Aug.).

Hampson, N. (1983), *Will and Circumstance: Montesquieu, Rousseau and the French Revolution*. Duckworth, London.

Hare, R. M. (1981), *Moral Thinking: Its Levels, Method and Point*. Clarendon Press, Oxford.

Hart, H. L. A. (1955), 'Are there any natural rights?' *Philosophical Review*, 64.

Hayek, F. A. (1960), *The Constitution of Liberty*. Routledge and Kegan Paul, London.

Hegel, G. W. F. (1822–31), *The Philosophy of History*, translated by J. Sibree, Introduction by C. J. Friedrich. Dover Publications, New York, 1956.

Heller, A. (1976), *The Theory of Need in Marx*. St Martin's Press, New York.

—— (1982), 'The Legacy of Marxian Ethics Today', *Praxis International*, 1, 4.

Holmstrom, N. (1977), 'Exploitation', *Canadian Journal of Philosophy*, VII. 2.

Howe, I. (1981), 'On the Moral Basis of Socialism', *Dissent*, (Fall).

—— (1982), *A Margin of Hope. An Intellectual Autobiography*. Secker and Warburg, London, 1983.

Hudson, W. (1982), *The Marxist Philosophy of Ernst Bloch*. Macmillan, London.

Hume, D. (1739), *A Treatise of Human Nature*. Ed. L. A. Selby-Bigge. Clarendon Press, Oxford, 1888, reprinted 1951.

Husami, Z. I. (1978), 'Marx on Distributive Justice', *Philosophy and Public Affairs*, 8. 1 (Fall) reprinted in Cohen, Nagel and Scanlon, 1980.

Jeanson, F. (1965), *Le Problème moral et la pensée de Sartre.* Editions de Seuil, Paris.

Kamenka, E. (1969), *Marxism and Ethics.* St. Martin's Press, New York.

—— (1972), *The Ethical Foundations of Marxism.* 2nd rev. edn., Routledge and Kegan Paul, London.

Kant, I. (1795), *Perpetual Peace. A Philosophical Essay.* Translated with Introduction and Notes by M. Campbell Smith. First published by Allen and Unwin, 1903, reprinted Garland Publishing Inc., New York and London, 1972.

Kautsky, K. (1906a), *Die Ethik und die materialistische Geschichts-auffassung.* Translated by J. R. Askew as *Ethics and the Materialistic Conception of History.* 4th rev. edn., Charles H. Kerr, Chicago, no date.

—— (1906b), 'Leben, Wissenschaft und Ethik', *Neue Zeit* XXIV. 2. (Reply to Bauer 1906.) Translated in part in Karl Kautsky, *Selected Political Writings*, edited and translated by P. Goode, Macmillan, London, 1983.

—— (1919), *Terrorismus und Kommunismus: Ein Beitrag zur Naturgeschichte der Revolution.* Berlin. Translated by W. H. Kerridge as *Terrorism and Communism*. Allen and Unwin, London. 1920.

Keck, T. R. (1975), 'Kant and Socialism. The Marburg School in Wilhelmian Germany'. Ph.D. thesis, History, University of Wisconsin.

Knei-Paz, B. (1978), *The Social and Political Thought of Leon Trotsky.* Oxford University Press, Oxford and New York.

Koestler, A. (1940), *Darkness at Noon*. Jonathan Cape: Penguin Books, London.

—— (1954), *The Invisible Writing: Autobiography 1931–53*. Collins with Hamish Hamilton, London.

Kolakowski, L. (1978), *Main Currents of Marxism*. 3 volumes. Clarendon Press, Oxford.

Kopelev, L. (1975), *No Jail for Thought*. Translated and edited by A. Austin. Penguin Books edn., Harmondsworth, 1979.

Korsch, K. (1930), *Der Gegenwärtige Stand des Problems 'Marxismus und Philosophie'*. Translated by Fred Halliday in *Marxism and Philosophy*. New Left Books, London, 1970.

Krieger, L. (1957), *The German Idea of Freedom. History of a Political Tradition*. University of Chicago Press, Chicago and London.

Kropotkin, P. (1885), *Paroles d'un revolté*. C. Marpon et E. Flammarion, Paris.

Lefort, C. (1981), *L' Invention démocratique: les limites de la domination totalitaire*. Fayard, Paris.

Lenin, V. I. (References are to the *Collected Works* in 45 volumes. Foreign Languages Publishing House, Moscow, 1960–3; Progress Publishers, 1964–70, indicated by CW.)

—— (1893), 'What the "Friends of the People" are and how they fight the Social Democrats', CW 1.

—— (1894), 'The Economic Content of Narodism', CW 1.

—— (1902), 'What is to be Done?', CW 5

—— (1915), 'Karl Marx', CW 21.

—— (1917a), 'The State and Revolution', CW 25.

—— (1917b), 'How to Organise Competition', CW 26.

—— (1918), Speech at the First Congress of Economic Councils, 26 May 1918, CW 27.

—— (1919), 'A Great Beginning', CW 29.

—— (1920a), 'From the Destruction of the Old Social System to the Creation of the New', CW 30.

—— (1920b), Speech at 3rd Komsomol Congress, 2 October 1920, CW 31.

Lukács, G. (1919a), 'Tactics and Ethics' Translated by Michael McColgan in *Political Writings: 1919–1929* edited by Rodney Livingstone. New Left Books, London, 1972.

—— (1919b), 'The Role of Morality in Communist Production', in ibid.

—— (1924), 'Bernstein's Triumph: Notes on the Essays written in honour of Karl Kautsky's Seventieth Birthday', in ibid.

Lukes, S. (1982), 'Can a Marxist believe in Human Rights?', *Praxis International*, 1.4.

Luxemburg, R. (1918), *The Russian Revolution*, in *The Russian Revolution* and *Leninism or Marxism?* with an introduction by B. D. Wolfe. University of Michigan Press, Ann Arbor, 1961.

McBride, W. L. (1977), *The Philosophy of Marx*. St Martin's Press, New York.

—— (1975), 'The Concept of Justice in Marx, Engels, and Others', *Ethics*, 85.

—— (1984), 'Rights and the Marxian Tradition', *Praxis International*, 4.1. (Comment on Lukes 1982.)

Macdonald, Dwight (1946), 'The Root is Man', *Politics*, 3. 4 (April).

Mackie, J. L. (1977), *Ethics: Inventing Right and Wrong*. Penguin, Harmondsworth.

Manuel, F. E. and Manuel, F. P. (1979), *Utopian Thought in the Western World*. Blackwell, Oxford.

Markovic, M. (1982), *Democratic Socialism: Theory and Practice*. Harvester, Brighton and St Martin's, New York.

—— and Petrovic, G. (eds.) (1969), *Praxis: Yugoslav Essays in the Philosophy and Methodology of the Social Sciences*. D. Reidel, Dordrecht.

Marx, K. (The *Collected Works* of Marx and Engels, published in English translation by Lawrence and Wishart, London, from 1975 are indicated by MECW; the *Selected Works* in two volumes published by the Foreign Languages Publishing House, Moscow, in 1962 by SW; and the *Selected Correspondence* published by the Foreign Languages Publishing House, Moscow (no date) by SC.)

—— (1843a), 'On the Jewish Question'. MECW 3.

—— (1843b), Letter to Ruge, May 1843. Translated in *The Letters of Karl Marx*. Selected and translated with explanatory notes by Saul K. Padover. Prentice-Hall, Englewood Cliffs, N. J., 1979

—— (1843c), Letter to Ruge, September 1843. Translated in ibid.

—— (1844a), 'Critique of Hegel's Philosophy of Law. Introduction', MECW 3.

—— (1844b), 'Comments on James Mill, *Éléments d'Économie Politique*'. MECW 3.

—— (1844c), *Economic and Philosophical Manuscripts*. MECW 3.

—— (1844d), Letter to Ludwig Feuerbach, 11 August 1844. MECW 3.

—— (1845), *Theses on Feuerbach*. MECW 5.

—— (1847a), *The Poverty of Philosophy*. MECW 6.

—— (1847b), 'The Communism of the *Rheinscher Beobachter*. MECW 6.

—— (1853), 'The British Rule in India'. SW 1.

—— (1857–8). *Grundrisse der Kritik der politischen Ökonomie*. Translated as *Foundations of the Critique of Political Economy (Rough Draft)* by Martin Nicolaus. Penguin Books in association with *New Left Review*, 1973.

Marx, K. (1859), Preface to *A Contribution to the Critique of Political Economy*. SW 1.

—— (1861–3), *Zur Kritik der Politischen Oekonomie* (Manuscript 1861–3), *Marx-Engels Gesamtausgabe*, Dietz, Berlin, 1982. (Notebooks VI to XV are included in Marx 1862–3, below.)

—— (1861–79), *Capital*, Vol. II, ed. Engels. Foreign Languages Publishing House, Moscow, 1957; and *Capital*, Vol III, ed. Engels. Foreign Langauges Publishing House, Moscow, 1962.

—— (1862–63), *Theories of Surplus Value*. Vol. 4 of *Capital*. Foreign Languages Publishing House, Moscow: Part 1, n.d.; Part 2, 1968; Part 3, 1972.

—— (1863–64), 'Results of the Immediate Process of Production'. Translated in *Capital*, Vol. I. Penguin Books in association with *New Left Review*, Harmondsworth, 1976.

—— (1864a), General Rules of the International Working Men's Association. SW 1.

—— (1864b), Inaugural Address of the Working Men's International Association. SW 1.

—— (1864c), Letter to F. Engels, 4 November 1864. SC.

—— (1865), *Wages, Price and Profit*. SW 1.

—— (1866), Letter to L. Kugelmann, 6 October 1866. SC.

—— (1867), *Capital*, Vol. I. Translation by Moore and Aveling, Foreign Languages Publishing House, Moscow, 1959.

—— (1871a), *The Civil War in France*. SW 1.

—— (1871b), First draft of *The Civil War in France*, *Marx-Engels Gesamtausgabe*, 1. 22. Dietz, Berlin, 1978.

—— (1875), *Critique of the Gotha Programme*. SW 2.

—— (1877), Letter to F. A. Sorge, 19 October 1877. SC.

——.(1879–80), 'Notes on Adolph Wagner'. Translated in Terrell Carver (ed.), *Karl Marx: Texts on Method*. Blackwell, Oxford, 1975.

—— (1880–2), *The Ethnological Notebooks of Karl Marx*. (Studies of Morgan, Phear, Maine, Lubbock.) Translated and edited with an introduction by L. Krader. Van Gorcum, Assen, 1972.

—— (1948), *Pages choisies pour une éthique socialiste*, ed. M. Rubel. Paris.

Marx, K. and Engels, F.

—— (1845), *The Holy Family*. MECW 4.

—— (1845–6), *The German Ideology*. MECW 5.

—— (1846), *Circular against Kriege*. MECW 6.

—— (1848), *Manifesto of the Communist Party*. MECW 6.

Medvedev, R. (1968), *Let History Judge*. Edited and translated by D. Joravsky, G. Haupt, and C. Taylor. Knopf, New York, 1971: Macmillan, London, 1972; and paperback edn., Spokesman Books, 1976.

Merleau-Ponty, M. (1947), *Humanisme et terreur. Essai sur le problème communiste*. Galimard, Paris (republished with an introduction by Claude Lefort, 1980). Translated with notes by John O'Neill as *Humanism and Terror. An Essay on the Communist Problem*. Beacon Press, Boston, 1969.

—— (1955), *Les Aventures de la dialectique*. Galimard, Paris. Translated by Joseph Bien as *Adventures of the Dialectic*. Heinemann, London, 1974.

Meszaros, I. (1980), 'Marxism and Human Rights', in A. D. Falconer (ed.), *Understanding Human Rights*, The Proceedings of the International Consultation held in Dublin 1978. Irish School of Economics, Dublin.

Miller, R. (1984), *Analyzing Marx*. Princeton University Press, Princeton.

Moore, Barrington, Jr. (1967), *Social Origins of Dictatorship and Democracy*. Allen Lane, The Penguin Press, London.

—— (1972), *Reflections on the Causes of Human Misery and upon Certain Proposals to Eliminate Them*. Beacon Press, Boston.

Moore, S. (1980), *Marx on the Choice between Socialism and Communism*. Harvard University Press, Cambridge, Mass. and London.

Nettl, J. P. (1966), *Rosa Luxemburg*. 2 vols. Oxford University Press, London.

Nielsen, K. and Patten, S. C. (eds.), (1981), *Marx and Morality*. Canadian Association for Publishing in Philosophy, Guelph, Ontario.

Nozick, R. (1974), *Anarchy, State and Utopia*. Basil Blackwell, Oxford.

Ollman, B. (1971), *Alienation*. Cambridge University Press, New York.

—— (1977), 'Marx's Vision of Communism', first published in *Critique*, Summer 1977, reprinted in Ollman, *Social and Sexual Revolution. Essays on Marx and Reich*. South End Press, Boston, 1979.

Pashukanis, E. B. (1924), *Law and Marxism. A General Theory*. Translated by Barbara Einhorn, edited and introduced by Chris Arthur. Ink Links, London, 1978.

Phillips, P. (1980), *Marx and Engels on Law and Laws*. Martin Robertson, Oxford.

Plamenatz, J. (1975), *Karl Marx's Philosophy of Man*. Oxford University Press, Oxford.

Plekhanov, G. V. (1903), Speech to Second Congress of the RSDLP, translated in part in Cliff 1975: 106.

—— (1908). *Fundamental Problems of Marxism*. Lawrence and Wishart, London; International Publishers, New York.

Praxis International, 1.1. (1981): Symposium on Socialism and Democracy.

158 Bibliography

Rawls, J. (1972), *A Theory of Justice*. Clarendon Press, Oxford.

Raz, J. (1975), *Practical Reason and Norms*. Hutchinson, London.

—— (1984), 'On the Nature of Rights', *Mind*, XCIII, 370 (April).

Renner, K. (1904), *The Institutions of Private Law and their Social Functions*. Translated by A. Schwarzchild and edited by O. Kahn-Freund. Routledge and Kegan Paul, London, 1949.

Roberts, J. M. (1966), *French Revolutionary Documents*, Volume I. Blackwell, Oxford.

Roemer, J. (1982), *A General Theory of Exploitation and Class*. Harvard University Press, Cambridge, Mass.

Salvadori, M. (1979), *Karl Kautsky and the Socialist Revolution*. New Left Books, London.

Sandel, M. (1982), *Liberalism and the Limits of Justice*. Cambridge University Press, Cambridge.

Sandkuehler, H. J. and De la Vega, R. (1970), *Marxismus und Ethik: Texte zum neokantischen Sozialismus*. Frankfurt.

Sandurski, W. (1983), 'To Each according to his (genuine?) Needs', *Political Theory*, 11. 3 (Aug.)

Sartre, J.-P. (1947–8), *Cahiers pour la morale*. Galimard, Paris, 1983.

—— (1948), *Les Mains sales*. Translated by Kitty Black as *Crime Passionel*. Methuen, London, 1961.

Scheffler, S. (1982), *The Rejection of Consequentialism: A Philosophical Investigation of the Considerations Underlying Rival Moral Conceptions*. Clarendon Press, Oxford.

Selucky, R. (1979), *Marxism, Socialism, Freedom*, Macmillan, London.

Sen, A. K. (1981), 'Rights and Agency', *Philosophy and Public Affairs*, 11. 1.

Serge, V. (1939a), 'Secrecy and Revolution—a Reply to Trotsky'. (Unknown whether or when originally published—probably dating from 1939.) Published in translation by Peter Sedgwick in *Peace News*, 27 Dec. 1963.

—— (1939b), Untitled Essay on Trotsky's *Their Morals and Ours*. (Undated—probably 1939.) Translated by P. Sedgwick (typescript).

—— (1942–43), *Memoirs of a Revolutionary*. Translated and edited by Peter Sedgwick. Oxford University Press, London, 1963.

Simmel, G. (1917), 'Individual and Society in Eighteenth- and Nineteenth-Century Views of Life. An Example of Philosophical Sociology' translated in *The Sociology of Georg Simmel*. Translated, edited, and with an introduction by K. H. Wolff. Free Press of Glencoe, Collier-Macmillan, London, 1950.

Skillen, A. (1974) 'Marxism and Morality', *Radical Philosophy*, 8.

—— (1977), *Ruling Illusions: Philosophy and the Social Order*. Harvester Press, Sussex.

Skocpol, T. (1979), *States and Social Revolutions*. Cambridge University Press, Cambridge.

Steinberg, H. J. (1967), *Sozialismus und deutsche Sozialdemokratie. Zur Ideologie der Partei vor dem 1 Weltkrieg.* Hannover.

Stojanovic, S. (1973), *Between Ideals and Reality: A Critique of Socialism and its Future.* Oxford University Press, New York.

Tadic, L. (1982), 'The Marxist Critique of Right in the Philosophy of Ernst Bloch', *Praxis International*, 1.4.

Taylor, C. (1979a), *Hegel and Modern Society.* Cambridge University Press, Cambridge.

—— (1979b), 'What's Wrong with Negative Liberty?', in A. Ryan (ed.), *The Idea of Freedom. Essays in honour of Isaiah Berlin.* Oxford University Press, Oxford.

Thompson, E. P. (1978), *The Poverty of Theory and Other Essays.* Merlin Press, London.

Trotsky, L. D. (1904), *Our Political Tasks.* New Park, London, 1980.

—— (1920), *Terrorism and Communism.* University of Michigan Press, Ann Arbor, 1961.

—— (1938), 'Their Morals and Ours', *The New International*, June 1938. Reproduced in *Their Morals and Ours. Marxist versus Liberal views on Morality.* Four essays by Leon Trotsky, John Dewey, and George Novack. 4th edn., Pathfinder Press, New York, 1969.

Tucker, R. C. (1969), *The Marxian Revolutionary Idea.* W. W. Norton, New York.

Vorländer, K. (1904), *Marx und Kant.* Vortrag genhalten in Wien am 8 April 1904. Verlag der "Deutschen Worte", Vienna.

—— (1911), *Kant und Marx. Beitrag zur Philosophie der Sozialismus.* 2nd edn., 1926. Tübingen.

Walicki, A. (1983), 'Marx and Freedom', *New York Review of Books*, xxx. 18.

Walzer, M. (1973), 'Political Action: The Problem of Dirty Hands', *Philosophy and Public Affairs*, 2.2 (Winter).

Willey, T. E. (1978), *Back to Kant. The Revival of Kantianism in German Social and Historical Thought 1860–1914.* Wayne State University Press, 1978.

Williams, B. (1978), 'Politics and Moral Character'; reprinted in *Moral Luck: Philosophical Papers 1973–1980.* Cambridge University Press, Cambridge.

Wood, A. (1972), 'The Marxian Critique of Justice', *Philosophy and Public Affairs*, 1. 3 (Spring), reprinted in Cohen, Nagel, and Scanlon 1980.

—— (1979), 'Marx on Right and Justice: A Reply to Husami', *Philosophy and Public Affairs*, 8. 3 (Spring), reprinted in ibid.

—— (1981), *Karl Marx.* Routledge and Kegan Paul, London.

Young, G. (1975–6), 'The Fundamental Contradiction of Capitalist Production', *Philosophy and Public Affairs*, 5.

Young, G. (1978), 'Justice and Capitalist Production: Marx and Bourgeois Ideology', *Canadian Journal of Philosophy*, 8.

—— (1981), 'Doing Marx Justice', in Nielsen and Patten 1981 (see above).

Index

MORE OXFORD PAPERBACKS

Details of a selection of other books follow. A complete list of Oxford Paperbacks, including The World's Classics, Twentieth-Century Classics, OPUS, Past Masters, Oxford Authors, Oxford Shakespeare, and Oxford Paperback Reference, is available in the UK from the General Publicity Department, Oxford University Press (JN), Walton Street, Oxford OX2 6DP.

In the USA, complete lists are available from the Paperbacks Marketing Manager, Oxford University Press, 200 Madison Avenue, New York, NY 10016.

Oxford Paperbacks are available from all good bookshops. In case of difficulty, customers in the UK can order direct from Oxford University Press Bookshop, 116 High Street, Oxford, Freepost, OX1 4BR, enclosing full payment. Please add 10 per cent of published price for postage and packing.

THREE ESSAYS

John Stuart Mill

On Liberty
Representative Government
The Subjection of Women

With an introduction by Richard Wollheim

The three major essays collected in this volume, written in the latter half of the life of John Stuart Mill (1806–73), were quickly accepted into the canon of European political and social thought. Nothing that has occurred in the intervening years has seriously affected their standing as classics on the subject. Today, although many of Mill's measures have been adopted, the essays are still relevant—when liberty and representative government are in collision with other principles, and when women still have to gain unprejudiced general acceptance of their equality.

In this introduction Richard Wollheim describes the essays as 'the distillation of the thinking of one highly intelligent, highly sensitive man, who spent the greater part of his life occupied with the theory and practice of society'.

MARXISM AND POLITICS

Ralph Miliband

Neither Marx nor any of his successors sought to define an overall theory of the nature of their political views. Professor Miliband has reconstructed from a wide range of material the main elements of the political theory and actual politics which are specific to Marxism. In so doing he highlights some of the problems left unresolved by earlier Marxists and discusses some pertinent questions of central importance to the politics of the twentieth century.

Marxist Introductions series

PERSONAL IMPRESSIONS

Isaiah Berlin

With an Introduction by Noel Annan

'an enthralling collection . . . It is hard to think of any other writer who is so penetrating, so amusing, and yet so entirely free of malice.' Anthony Storr in *Spectator*

This enthusiastically received collection contains Isaiah Berlin's appreciations of thirteen men of unusual distinction in the intellectual and political world—sometimes in both. The names of many of them are familiar—Churchill, Roosevelt, Weizmann, Namier, Austin, Bowra, Einstein. With the exception of Roosevelt he met them all, and he knew many of them well. The volume is completed by a vivid and moving account of his meetings in Russia with Boris Pasternak and Anna Akhmatova in 1945 and 1956. As Alan Ryan wrote in the *Sunday Times*, 'This last essay, in particular, is simply stunning.'

'This is a very moving and serious book, as well as a delightful one.' Richard Cobb in *Guardian*

MEN OF IDEAS

Some Creators of Contemporary Philosophy

Bryan Magee

In his successful BBC TV series 'Men of Ideas' Bryan Magee came face to face with fifteen of the world's foremost philosophers. The resulting discussions, edited transcripts of which appear in this book, add up to a lively yet authoritative introduction to much of the influential philosophy of our time.

'No clearer or more stimulating work on the creators of contemporary philosophy has appeared since Bertrand Russell's *History of Western Philosophy*.' *Books and Bookmen*

KARL MARX: HIS LIFE AND ENVIRONMENT

Fourth edition

Isaiah Berlin

Isaiah Berlin's brilliant account of the life and doctrines of the author of *Das Kapital* has long been established as a classic of intellectual biography. It provides a lucid, comprehensive introduction to the traditional Marx—his personality and ideas as they were understood by those who, in his name and guided by his ideas, made the revolutions which transformed the world. For this new edition the author has written a new preface, revised the text throughout, and added a number of fresh passages.

'As a portrait of the man and the intellectual climate of the mid-nineteenth century it is, perhaps, the finest we have in any European language.' Chimen Abramsky

An OPUS book

MARX'S SOCIAL THEORY

Terrell Carver

Why has Marx had such a wide-ranging impact on our intellectual and political life? Terrell Carver presents a new analysis of what Marx called the 'guiding thread' of his studies, which is set out in his 1859 preface *A Critique of Political Economy*, together with an important autobiographical sketch, which the author reanalyses in this book. He argues that Marx's 'production theory of society and social change' is analogous to Darwin's work in a hitherto unnoticed way and is just as scientific. He assesses the central difficulties encountered by the theory, and shows that it sprang from a desire not simply to interpret the world, but to change it.

THE PROBLEMS OF PHILOSOPHY
Bertrand Russell

First published in 1912, this classic introduction to the subject of philosophical inquiry has proved invaluable to the formal student and general reader alike. It has Russell's views succinctly stated on material reality and idealism, knowledge by acquaintance and by description, induction, knowledge of general principles and of universals, intuitive knowledge, truth and falsehood, the distinctions between knowledge, error, and probable opinion, and the limits and the value of philosophical knowledge.

A foreword Russell wrote in 1924 for a German translation has been added as an appendix. Here Russell gave details of how some of his views had changed since *The Problems of Philosophy* was written.

An OPUS book

FOUR ESSAYS ON LIBERTY
Isaiah Berlin

The four essays are 'Political Ideas in the Twentieth Century'; 'Historical Inevitability'; 'Two Concepts of Liberty', a ringing manifesto for pluralism and individual freedom; and 'John Stuart Mill and the Ends of Life'. There is also a long and masterly Introduction written specially for this collection, in which the author replies to his critics.

'The densely-written, richly allusive style perfectly matches the contents; practically every paragraph introduces us to half a dozen new ideas and as many thinkers—the landscape flashes past, peopled with familiar and unfamiliar people, all arguing incessantly.' *New Society*

THE AGE OF ENLIGHTENMENT
The Eighteenth-Century Philosophers
Isaiah Berlin

'The intellectual power, honesty, lucidity, courage and disinterested love of the truth of the most gifted thinkers of the eighteenth century remain to this day without parallel. Their age is one of the best and most hopeful episodes in the life of mankind.'

These are the closing words of Isaiah Berlin's introduction to his selection, with running commentary, from the major works of Locke, Berkeley, Hume and other leading eighteenth-century philosophers. This book provides an excellent starting-point for the study of the Enlightenment, letting the leading lights of the period speak in their own (readable) words, and providing background information and elucidation where necessary.

MAIN CURRENTS OF MARXISM
Leszek Kolakowski
Volume 3: The Breakdown

Leszek Kolakowski begins this third volume with an analysis of Stalinism and a discussion of the impact of Marxism on Soviet Culture. He examines the contributions of Trotsky, Gramsci, Lukacs, Marcuse, and others, and traces the developments in Marxism since the Second World War.

'Undoubtedly the most complete and intellectually satisfying survey of Marx's and Marxist thought ever written.'   *Listener*

'must be acknowledged as opening a new era in Marxist criticism'   *American Scholar*

MAIN CURRENTS OF MARXISM

Leszek Kolakowski

Volume 1: The Founders

In this first volume, Leszek Kolakowski examines the origins of Marxism, tracing its descent, from the neo-Platonists through Hegel and the Enlightenment. He analyses the development of Marx's thought and shows its divergence from other forms of socialism.

'the most commanding, the most decisive, the most properly passionate and yet also . . . the most accessible account of Marxism that we now have. It is a work of surpassing lucidity and power, of the sharpest and most sensitive judgement, of a far finer quality than almost all of that with which it deals. It is, in short, a masterpiece.'

Times Higher Educational Supplement

MARX

Peter Singer

Peter Singer identifies the central vision that unifies Marx's thought, enabling us to grasp Marx's views as a whole. He views him as a philosopher primarily concerned with human freedom, rather than as an economist or social scientist. He explains alienation, historical materialism, the economic theory of *Capital,* and Marx's idea of communism, in plain English, and concludes with a balanced assessment of Marx's achievement.

'an admirably balanced portrait of the man and his achievement'
Observer

Past Masters

THE GREAT PHILOSOPHERS
From Plato to the Present Day

Bryan Magee

Beginning with the death of Socrates in 399, and following the story through the centuries to recent figures such as Bertrand Russell and Wittgenstein, Bryan Magee and fifteen contemporary writers and philosophers provide an accessible and exciting introduction to Western philosophy and its greatest thinkers.

'Magee is to be congratulated . . . anyone who sees the programmes or reads the book will be left in no danger of believing philosophical thinking is unpractical and uninteresting.' Ronald Hayman, *Times Educational Supplement*

'one of the liveliest, fast-paced introductions to philosophy, ancient and modern, that one could wish for' *Universe*

HEGEL

Peter Singer

Many people regard Hegel's work as obscure and extremely difficult, yet his importance and influence are universally acknowledged. Professor Singer eliminates any excuse for remaining ignorant of the outlines of Hegel's philosophy by providing a broad discussion of his ideas, and an account of his major works.

'an excellent introduction to Hegel's thought . . . Hegel is neatly placed in historical context; the formal waltz of dialectic and the dialectic of master and slave are economically illumined; Singer's use of analogy is at times inspired.' *Sunday Times*

Past Masters

MAIN CURRENTS OF MARXISM

Leszek Kolakowski

Volume 2: The Golden Age

In this second volume, Leszek Kolakowski examines the doctrine of the leading Marxists, and the controversies between them, in the era of the Second International. At this time there were many interesting varieties of Marxism, and the leading contributors to the debate are all discussed—Kautsky, Rosa Luxemburg, Plekhanov, Lenin, and Georges Sorel among them.

'remarkable for almost more reasons than one could list . . . one thing that cannot be over-emphasised is the sheer usefulness of the book: the second volume, in particular, is an invaluable scholarly guide to the richness of Marxism in "the Golden Age" . . . a distinctive and engrossing piece of work'
Times Educational Supplement

MARXISM AND LITERATURE

Raymond Williams

This book extends the theme of Raymond Williams's earlier work in literary and cultural analysis. He analyses previous contributions to a Marxist theory of literature from Marx himself to Lukács, Althusser, and Goldmann, and develops his own approach by outlining a theory of 'cultural materialism' which integrates Marxist theories of language with Marxist theories of literature.

'Williams has brought his authority and experience, established by his immense critical achievement, into the Marxist tradition.' Anthony Barnett, *New Society*

Marxist Introduction series

MARXISM AND PHILOSOPHY

Alex Callinicos

Marxism began with the repudiation of philosophy. Marx declared: 'The philosophers have only *interpreted* the world in various ways; the point is to *change* it.' Yet Marxists have often resorted to philosophical modes of reasoning, and Western Marxism has recently been more concerned with philosophy than with empirical research or political activity.

This book explores the ambivalent attitude of Marxism to philosophy, starting with an examination of the Marxist view of Hegel. Alex Callinicos goes on to contrast the German classical idealism, from which Marxist philosophy stems, with the very different tradition of analytical philosophy prevalent among English-speaking philosophers.

Marxist Introductions series

MARXISM AND LAW

Hugh Collins

In this introduction to Marxism and the law, Hugh Collins presents a unified and coherent view of Marxism, which he uses to examine the specific characteristics of legal institutions, rules, and ideals. He pays particular attention to the place of ideology in law, the distinction between base and superstructure, and the destiny of law in a Communist society, and frequently subjects the Marxist approach to criticism, suggesting that many of the Marxist claims about law are unproven or misconceived.

'enthusiastically and warmly recommended' *New Law Journal*

Marxist Introduction series

ENGELS

Terrell Carver

In a sense, Engels invented Marxism. His chief intellectual legacy, the materialist interpretation of history, has had a revolutionary effect on the arts and social sciences, and his work as a whole did more than Marx's to make converts to the most influential political movement of modern times. In this book Terrell Carver traces Engel's career, and looks at the effect of the materialist interpretation of history on Marxist theory and practice.

'Carver's refreshingly honest book . . . is packed with careful judgements about the different contributions of Engels to 19th century marxism.' *New Society*

Past Masters

MODERN BRITISH PHILOSOPHY

Bryan Magee

Bryan Magee's general survey of contemporary British philosophy originates in a series of radio conversations with leading British philosophers. In them, he elicited their views on chosen subjects, ranging from influential philosophers, such as Russell, Wittgenstein, Moore, and Austin, to the relationship of philosophy to Morals, Religion, the Arts, and Social Theory.

'Under Magee's sensitive guidance a remarkably coherent interpretation of this period emerges.' Marshall Cohen in *The Listener*

'Bryan Magee has done everything possible to get these philosophers to put themselves in a nutshell.' Kathleen Nott in *New Society*